Note Taking Guide

Precalculus
Mathematics for Calculus

SIXTH EDITION

James Stewart
McMaster University and University of Toronto

Lothar Redlin
The Pennsylvania State University

Saleem Watson
California State University, Long Beach

Prepared by

Emily J. Keaton

BROOKS/COLE
CENGAGE Learning™

Australia • Brazil • Japan • Korea • Mexico • Singapore • Spain • United Kingdom • United States

BROOKS/COLE
CENGAGE Learning™

For product information and technology assistance, contact us at
**Cengage Learning Customer & Sales Support,
1-800-354-9706**

For permission to use material from this text or product, submit all requests online at **www.cengage.com/permissions**
Further permissions questions can be emailed to
permissionrequest@cengage.com

ISBN-13: 978-1-111-57259-4
ISBN-10: 1-111-57259-3

Brooks/Cole
20 Davis Drive
Belmont, CA 94002-3098
USA

Cengage Learning is a leading provider of customized learning solutions with office locations around the globe including Singapore, the United Kingdom, Australia, Mexico, Brazil, and Japan. Locate your local office at: **www.cengage.com/global**

Cengage Learning products are represented in Canada by Nelson Education, Ltd.

To learn more about Brooks/Cole, visit
www.cengage.com/brookscole

Purchase any of our products at your local college store or at our preferred online store
www.cengagebrain.com

Printed in the United States of America
2 3 4 5 6 7 15 14 13 12

Table of Contents

Chapter 1   Fundamentals

1.1   Real Numbers

Give examples or descriptions of the types of numbers that make up the real number system.

Natural numbers: _____

Integers: _____

Rational numbers: _____

Irrational numbers: _____

The set of all real numbers is usually denoted by the symbol _____.

The corresponding decimal representation of a rational number is _____. The
corresponding decimal representation of an irrational number is _____.

I. Properties of Real Numbers

Let a, b, and c be any real numbers. Use a, b, and c to write an example of each of the following properties of real numbers.

Commutative Property of Addition: _____

Commutative Property of Multiplication: _____

Associative Property of Addition: _____

Associative Property of Multiplication: _____

Distributive Properties: _____

_____

Example 1:   Use the properties of real numbers to write $4(q+r)$ without parentheses.

II. Addition and Subtraction

The **additive identity** is _____ because, for any real number a, _____. Every real
number a has a _____ , $-a$, that satisfies $a+(-a)=$ _____.

To subtract one number from another, simply _____.

Complete the following Properties of Negatives.

1. $(-1)a=$ _____

2. $-(-a) =$ _____

3. $(-a)b =$ _____

4. $(-a)(-b) =$ _____

5. $-(a + b) =$ _____

6. $-(a - b) =$ _____

Example 2:   Use the properties of real numbers to write $-3(2a - 5b)$ without parentheses.

III. Multiplication and Division

The **multiplicative identity** is _____ because, for any real number a, _____. Every nonzero real number a has an _____ , $1/a$, that satisfies $a \cdot (1/a) =$ _____.

To divide by a number, simply _____.

Complete the following Properties of Fractions.

1. $\dfrac{a}{b} \cdot \dfrac{c}{d} =$ _____

2. $\dfrac{a}{b} \div \dfrac{c}{d} =$ _____

3. $\dfrac{a}{c} + \dfrac{b}{c} =$ _____

4. $\dfrac{a}{b} + \dfrac{c}{d} =$ _____

5. $\dfrac{ac}{bc} =$ _____

6. If $\dfrac{a}{b} = \dfrac{c}{d}$, then _____

Example 3:   Evaluate: $\dfrac{4}{9} + \dfrac{19}{30}$

IV. The Real Line

On the real number line shown below, the point corresponding to the real number 0 is called the

_____. Given any convenient unit of measurement, each positive number x is represented by

_____. Each negative

number $-x$ is represented by _____ .

The real numbers are *ordered,* meaning that **a is less than b,** written _____, if _____

_____. The symbol $a \le b$ is read as _____.

V. Sets and Intervals

A **set** is a _____, and these objects are called the _____ of

the set. The symbol $\in$ means _____, and the symbol $\notin$ means _____

_____.

Name two ways that can be used to describe a set.

The **union** of two sets S and T is the set $S \cup T$ that consists of _____

_____. The **intersection** of S and T is the set $S \cap T$ that consists of _____

_____. The symbol $\varnothing$ represents _____

_____.

Example 4:   If $A = \{2,4,6,8,10\}$, $B = \{4,8,12,16\}$, and $C = \{3,5,7\}$, find the sets
     (a) $A \cup B$
     (b) $A \cap B$
     (c) $B \cap C$

If $a < b$, then the **open interval** from a to b consists of _____ and

is denoted _____. The **closed interval** from a to b includes _____

and is denoted _____.

VI. Absolute Value and Distance

The **absolute value** of a number a, denoted by _____, is _____

_____. Distance is always _____, so we have

$|a| \geq 0$ for every number a.

If a is a real number, then the **absolute value** of a is

$$|a| = \left\{\right.$$

Example 5:   Evaluate.
          (a) $|12-8|$
          (b) $|9-15|$
          (c) $|7-7|$

Complete the following descriptions of properties of absolute value.

1. The absolute value of a number is always _____.

2. A number and its negative have the same _____.

3. The absolute value of a product is _____.

4. The absolute value of a quotient is _____.

If a and b are real numbers, then the **distance** between the points a and b on the real line is

$d(a,b) =$ _____.

Example 6:   Find the distance between the numbers -16 and 7.

Homework Assignment

Page(s)

Exercises

Name _____ Date _____

1.2 Exponents and Radicals

I. Integer Exponents

If a is any real number and n is a positive integer, then the _____ is $a^n = \underbrace{a \cdot a \cdots \cdot a}_{n \text{ factors}}$.

The number a is called the _____, and n is called the _____.

If $a \neq 0$ is any real number and n is a positive integer, then $a^0 = $ _____ and $a^{-n} = $ _____.

Example 1: Evaluate.

(a) $(-2)^5$ (b) $\left(\dfrac{1}{9}\right)^0$ (c) 4^{-2}

II. Rules for Working with Exponents

Complete the following Laws of Exponents.

1. $a^m a^n = $ _____

2. $\dfrac{a^m}{a^n} = $ _____

3. $(a^m)^n = $ _____

4. $(ab)^n = $ _____

5. $\left(\dfrac{a}{b}\right)^n = $ _____

6. $\left(\dfrac{a}{b}\right)^{-n} = $ _____

7. $\dfrac{a^{-n}}{b^{-m}} = $ _____

Example 2: Evaluate.

(a) $y^6 y^8$ (b) $(w^5)^3$ (c) $\left(\dfrac{b}{2}\right)^{-3}$

III. Scientific Notation

Scientists use exponential notation as a compact way of writing _____

_____ .

A positive number x is said to be written in _____ if it is expressed as $x = a \times 10^n$, where $1 \le a < 10$ and n is an integer.

Example 3:   Write each number in scientific notation.
(a) 1,750,000   (b) 0.0000000429

IV. Radicals

The symbol $\sqrt{}$ means _____ .

If n is any positive integer, then the **principal nth root** of a is defined as follows:

$$\sqrt[n]{a} = b \quad \text{means} \quad \underline{\hspace{3cm}} .$$

If n is even, we must have _____ .

Complete the Properties of nth Roots.

1. $\sqrt[n]{ab} = $ _____

2. $\sqrt[n]{\dfrac{a}{b}} = $ _____ _____

3. $\sqrt[m]{\sqrt[n]{a}} = $ _____ _____

4. $\sqrt[n]{a^n} = $ _____

5. $\sqrt[n]{a^n} = $ _____

Example 4:   Evaluate: $\sqrt[4]{64 w^5 y^8}$

V. Rational Exponents

For any rational exponent m/n in lowest terms, where m and n are integers and $n > 0$, we define

$a^{m/n} = $ _____   or equivalently $a^{m/n} = $ _____

If n is even, then we require that _____.

Example 5:   Evaluate.

a) $b^{7/8} b^{9/8}$

b) $\sqrt[4]{x^2 (x^5)^2}$

VI. Rationalizing the Denominator

Rationalizing the denominator is the procedure in which _____
_____.

Describe a strategy for rationalizing a denominator.

Example 6:   Rationalize the denominator: $\dfrac{x}{\sqrt{3y}}$

Additional notes

Homework Assignment

Page(s)

Exercises

Name _____ Date _____

1.3 Algebraic Expressions

A **variable** is _____.

An **algebraic expression** is _____
_____.

A **monomial** is _____
_____.

A **binomial** is _____.

A **trinomial** is _____.

A **polynomial** in the variable x is an expression of the form _____,

where $a_0, a_1, \ldots, a_n$ are real numbers, and n is a nonnegative integer. If $a_n \neq 0$, then the polynomial has

degree _____. The monomials $a_k x^k$ that make up the polynomial are called the

_____ of the polynomial.

The degree of a polynomial is _____
_____.

I. Adding and Subtracting Polynomials

We **add** and **subtract** polynomials by _____
_____. The idea is to combine _____, which are

terms with the same variables raised to the same powers, using the _____.

When subtracting polynomials, remember that if a minus sign precedes an expression in parentheses, then _____
_____.

II. Multiplying Algebraic Expressions

Explain how to find the **product** of polynomials or other algebraic expressions.

Explain the acronym FOIL.

Example 1:   Multiply: $(x-5)(3x+7)$

III. Special Product Formulas

Complete the following Special Product Formulas.

Sum and Difference of Same Terms

$(A + B)(A - B) =$ _____

Square of a Sum and Difference

$(A + B)^2 =$ _____

$(A - B)^2 =$ _____

Cube of a Sum and Difference

$(A + B)^3 =$ _____

$(A - B)^3 =$ _____

The key idea in using these formulas is the **Principle of Substitution,** which says that _____

_____.

Example 2:   Find the product: $(2y-5)^2$.

IV. Factoring Common Factors

Factoring an expression means _____

_____.

Example 3: Factor: $14x^3 - 2x^2$

V. Factoring Trinomials

To factor a trinomial of the form $x^2 + bx + c$, we note that $(x + r)(x + s) = x^2 + (r + s)x + rs$ so we need to choose numbers r and s so that _____.

To factor a trinomial of the form $ax^2 + bx + c$ with $a \neq 1$, we look for factors of the form $px + r$ and $qx + s$:

$ax^2 + bx + c = (px + r)(qx + s) = pqx^2 + (ps + qr)x + rs$. Therefore, we try to find numbers p, q, r, and s such that _____.

Example 4: Factor: $6x^2 + 7x - 3$

VI. Special Factoring Formulas

Complete the following Special Factoring Formulas.

Difference of Squares

$A^2 - B^2 = $ _____

Perfect Squares

$A^2 + 2AB + B^2 = $ _____

$A^2 - 2AB + B^2 = $ _____

Difference and Sum of Cubes

$A^3 - B^3 = $ _____

$A^3 + B^3 = $ _____

Example 5:   Factor:
   (a) $36 - 25x^2$ (b) $49x^2 + 28xy + 4y^2$

Describe how to recognize a perfect square trinomial.

VII. Factoring by Grouping Terms

Polynomials with at least four terms can sometimes be factored by _____.

Homework Assignment

Page(s)

Exercises

Name _____ Date _____

1.4   Rational Expressions

A **fractional expression** is _____.

A **rational expression** is _____
_____.

I. The Domain of an Algebraic Expression

The **domain** of an algebraic expression is _____
_____.

Example 1:   Find the domain of the expression $\dfrac{2x}{x^2+6x+5}$.

II. Simplifying Rational Expressions

Explain how to simplify rational expressions.

Example 2:   Simplify: $\dfrac{2x+2}{x^2+6x+5}$

III. Multiplying and Dividing Rational Expressions

To **multiply rational expressions,** use the following property of fractions:

$$\frac{A}{B} \cdot \frac{C}{D} = \underline{\hspace{5cm}}$$

This says that to multiply two fractions, we _____

_____ .

To **divide rational expressions,** use the following property of fractions:

$$\frac{A}{B} \div \frac{C}{D} = \underline{\hspace{5cm}}$$

This says that to divide a fraction by another fraction, we _____

_____ .

Example 3:   Perform the indicated operation and simplify: $\dfrac{x^2-1}{x+3} \div \dfrac{x^2+2x+1}{x^2-9}$.

IV. Adding and Subtracting Rational Expressions

To **add or subtract rational expressions,** we first find a common denominator and then use the following property of fractions:

$$\frac{A}{C} + \frac{B}{C} = \underline{\hspace{5cm}}$$

It is best to use the **least common denominator (LCD),** which is found by _____

_____

_____ .

Example 4:   Perform the indicated operation and simplify: $\dfrac{x-1}{x+3} + \dfrac{x^2+2x+1}{x^2-9}$

V. Compound Fractions

A **compound fraction** is _____

_____ .

Describe two different approaches to simplifying a compound fraction.

VI. Rationalizing the Denominator or the Numerator

If a fraction has a denominator of the form $A + B\sqrt{C}$, describe how to rationalize the denominator.

If a fraction has the numerator $\sqrt{3y} - 2$, how would you go about rationalizing the numerator?

VII. Avoiding Common Errors

Identify the error in the following solution and show the correct solution.

$$\frac{3}{2y} + \frac{2}{3y} = \frac{3+2}{2y+3y} = \frac{5}{5y} = \frac{1}{y}$$

Additional notes

Homework Assignment

Page(s)

Exercises

Name _____ Date _____

1.5 Equations

An equation is _____.

The **solutions** or **roots** of an equation are _____

_____. The process of finding these solutions is called

___solving the eq._____.

Two equations with exactly the same solutions are called _____.

Describe how to solve an equation.

Give a description of each property of equality.

1. $A = B \iff A + C = B + C$

   ___adding the same to both side___

   _____   _____

2. $A = B \iff CA = CB \quad (C \neq 0)$

   ___multiplying the same to both sides___

   _____   _____

I. Solving Linear Equations

The simplest type of equation is a *linear equation,* or ___first degree___, which is

an ___eq. in which each term is either a___

___constant or non zero multiple to variable.___

A **linear equation** in one variable is an equation equivalent to one of the form ___$ax + b = 0$___, where *a*

and *b* are real numbers and *x* is the variable.

Example 1: Solve the equation $5x + 8 = 2x - 7$.

$$5x + 8 = 2x - 7$$
$$-5x + 7 \qquad -5x + 7$$
$$\frac{15}{-3} = \frac{-3x}{-3} = \qquad -5 = x \checkmark$$

Example 2:   Solve for the variable b in the equation $A = h\left(\dfrac{a+b}{2}\right)$.

$$\left(A = h\left[\dfrac{a+b}{2}\right]\right)^2$$

$$\dfrac{A^2}{2h} = \dfrac{h(a+b) \cdot 2}{h \qquad 2}$$

$$\dfrac{A^2}{2h} - (a+b)$$

II. Solving Quadratic Equations

Quadratic equations are _____ 2nd _____ degree equations.

A **quadratic equation** is an equation of the form _____ $ax^2 + bx + c = 0$ _____, where a, b, and c are real numbers with $a \neq 0$.

The **Zero-Product Property** says that _____ $AB = 0$ if o if $A = 0$ or $B = 0$.

This means that if we can factor the left-hand side of a quadratic (or other) equation, then we can solve it by _____ turning equal to zero. _____.

Example 3:   Solve the equation $x^2 - 7x = 44$.

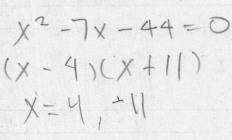

$$x^2 - 7x - 44 = 0$$

$$(x - 4)(x + 11)$$

$$x = 4, \ ^+11$$

$$\begin{array}{c} 44 \\ +4 \bcancel{X} \cancel{-11} \\ \cancel{7} \\ -4 \\ 11 \end{array}$$

The solutions of the simple quadratic equation $x^2 = c$ are _____ $X = 4 \quad X = -1$ _____.

If a quadratic equation does not factor readily, then we can solve it using the technique of _____ comple the sq. _____.

In this technique, to make $x^2 + bx$ a perfect square, add _____ $\left(b/2\right)^2$ of half the coefficient _____. This gives the perfect square $x^2 + bx \left(\dfrac{b}{2}\right)^2 = \left(X + \dfrac{b}{2}\right)^2$

State the **Quadratic Formula.**

$$\frac{-b \pm \sqrt{b^2 - 4ac}}{2a}$$

Example 4: Find all solutions of the equation $2x^2 - 3x - 6 = 0$.

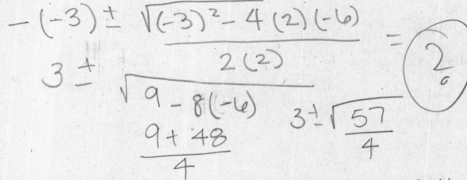

$$\frac{A \quad B \quad C}{(2)(-3)(-6)}$$

$$\frac{-(-3) \pm \sqrt{(-3)^2 - 4(2)(-6)}}{2(2)} = \boxed{\frac{2}{0}}$$

$$3 \pm \sqrt{\frac{9 - 8(-6)}{4}}$$

$$\frac{9 + 48}{4} \qquad \frac{3 \pm \sqrt{57}}{4}$$

The **discriminant** of the general quadratic equation $ax^2 + bx + c = 0$, $(a \neq 0)$, is $\underline{b^2 - 4ac}$.

1. If $D > 0$, then $\underline{2 \text{ distincs sols. real}}$.
2. If $D = 0$, then $\underline{\text{exactly 1 sol real}}$.
3. If $D < 0$, then $\underline{\text{no real sols.}}$.

Example 5: Use the discriminant to determine how many real solutions the equation $5x^2 - 16x + 4 = 0$ has.

$$\overset{a \quad b \quad c}{5x^2 - 16x + 4 = 0}$$

$$b^2 - 4ac$$

$$(-16)^2 - 4(5)(4)$$

$$-4(20) \qquad -386$$

$$-80 \qquad \text{no real} \qquad \boxed{2}$$

III. Other Types of Equations

When you solve an equation that involves radicals, you must be especially careful to _____

_____ because you may end up with one or more extraneous solutions, which

are _____.

An equation of **quadratic type** is an equation of the form _____

_____.

Example 6:   Find all solutions of the equation $x^4 - 52x^2 + 576 = 0$.

Additional notes

Name _____ Date _____

1.6   Modeling with Equations

I. Making and Using Models

List and explain the Guidelines for Modeling with Equations.

1.

2.

3.

4.

II. Problems About Interest

What is **interest?**

The most basic type of interest is _____, which is just an annual percentage of

_____.

The amount of a loan or deposit is called the _____. The annual percentage paid for the use of this money is the _____. The variable t stands for _____ _____, and the variable I stands for the _____ _____.

The **simple interest formula** gives the amount of interest I earned when a principal P is deposited for t years at an interest rate r and is given by _____. When using this formula, remember to convert r from a percentage to _____.

Example 1:   Consider the following situation:
Oliver deposits $22,000 at a simple interest rate of 4.25%. How much interest will he earn after 8 years?
In this situation, identify the value of each variable in the simple interest formula and indicate which variable is unknown.

III. Problems About Area or Length

Give formulas for the (1) area A and (2) perimeter P of a rectangle having length l and width w.

Give formulas for the (1) area A and (2) perimeter P of a square having sides of length s.

Give formulas for the (1) area A and (2) perimeter P of the given triangle.

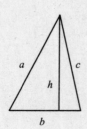

IV. Problems About Mixtures

Problems involving mixtures and concentrations make use of the fact that if an amount x of a substance is dissolved in a solution with volume V, then the concentration C of the substance is given by

_____.

Solving a mixture problem usually requires analyzing _____
_____.

V. Problems About the Time Needed to Do a Job

When solving a problem that involves determining how long it takes several workers to complete a job, use the fact that if a person or machine takes H time units to complete the task, then in one time unit the fraction of the task that has been completed is _____.

If it takes Faith 80 minutes to complete a task, what fraction of the task does she complete in *one hour*?

VI. Problems About Distance, Rate, and Time

Give the formula that relates the distance traveled by an object traveling at either a constant or average speed in a given amount of time.

Give an example of an application problem that requires this formula.

Additional notes

Homework Assignment

Page(s)

Exercises

Name _____ Date _____

1.7   Inequalities

An **inequality** looks _____
_____.

To **solve** an inequality that contains a variable means _____
_____. Unlike an equation, an inequality
generally has _____ solutions, which form _____
_____.

Describe how to solve an inequality.

Give a description of each property of inequality.

1. $A \leq B \iff A + C \leq B + C$

    _____
    _____   _____

2. $A \leq B \iff A - C \leq B - C$

    _____
    _____   _____

3. If $C > 0$, then $A \leq B \iff CA \leq CB$

    _____
    _____   _____

4. If $C < 0$, then $A \leq B \iff CA \geq CB$

    _____
    _____   _____

5. If $A > 0$ and $B > 0$, then $A \leq B \iff \dfrac{1}{A} \geq \dfrac{1}{B}$

    _____
    _____   _____

6. If $A \leq B$ and $C \leq D$, then $A + C \leq B + D$

_____

_____   _____

I. Solving Linear Inequalities

An inequality is **linear** if _____

_____. To solve a linear inequality, _____

_____.

Example 1:   Solve the inequality $5x + 9 \geq 2x - 9$.

II. Solving Nonlinear Inequalities

If a product or a quotient has an *even* number of negative factors, then its value is _____.

If a product or a quotient has an *odd* number of negative factors, then its value is _____.

State the Guidelines for Solving Nonlinear Inequalities.

1.

2.

3.

4.

5.

Example 2:   Solve the inequality $x^2 \leq 7x + 44$.

III. Absolute Value Inequalities

For each absolute value inequality, write an equivalent form.

1. $|x| < c$ $\Leftrightarrow$ _____

2. $|x| \leq c$ $\Leftrightarrow$ _____

3. $|x| > c$ $\Leftrightarrow$ _____

4. $|x| \geq c$ $\Leftrightarrow$ _____

Example 3:   Solve the inequality $|4x - 2| \leq 10$.

III. Modeling with Inequalities

Give an example of a real-life problem that can be solved with inequalities.

Additional notes

Homework Assignment

Page(s)

Exercises

Name _____ Date _____

1.8 Coordinate Geometry

I. The Coordinate Plane

Just as points on a line can be identified with real numbers to form the coordinate line, points in a plane can be

identified with ordered pairs of numbers to form the _____ or _____

_____.

Describe how the coordinate plane is constructed. Include a description of its major components: **x-axis, y-axis, origin,** and **quadrants**.

On the coordinate plane shown below, label the x-axis, the y-axis, the origin, and Quadrants I, II, III, and IV.

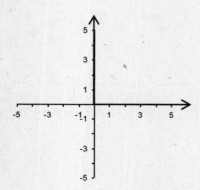

Any point P in the coordinate plane can be located by a unique ordered pair of numbers (a, b), where the first

number a is called the _____, and the second number b is called the

_____.

II. The Distance and Midpoint Formulas

The distance between the points $A(x_1, y_1)$ and $B(x_2, y_2)$ in the plane is given by

_____

Example 1:   Find the distance between the points $(-5, 6)$ and $(3, -5)$.

The midpoint of the line segment from $A(x_1, y_1)$ to $B(x_2, y_2)$ is

_____

Example 2:   Find the midpoint of the line segment from $(-5, 6)$ to $(3, -5)$.

III. Graphs of Equations in Two Variables

The **graph** of an equation in x and y is _____

_____.

Explain how to graph an equation.

Example 3: Sketch the graph of the equation $\frac{1}{2}x + y = 3$.

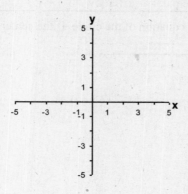

IV. Intercepts

Define x-intercepts.

Explain how to find x-intercepts.

Define y-intercepts.

Explain how to find y-intercepts.

Example 4: Find the x- and y-intercepts of the graph of the equation $y = 4x - 36$.

V. Circles

An equation of the circle with center (h, k) and radius r is _____ .

This is called the _____ for the equation of the circle. If the center of the circle

is the origin $(0, 0)$, then the equation is _____ .

Example 5:   Graph the equation $(x+2)^2 + (y-1)^2 = 4$.

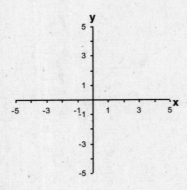

VI. Symmetry

A graph is **symmetric with respect to the y-axis** if whenever the point (x, y) is on the graph, then so is

_____ . To test an equation for this type of symmetry, replace _____ by

_____ . If the equation is unchanged, then the graph is symmetric with respect to the y-axis.

A graph is **symmetric with respect to the x-axis** if, whenever the point (x, y) is on the graph, then so is

_____ . To test an equation for this type of symmetry, replace _____ by

_____ . If the equation is unchanged, then the graph is symmetric with respect to the x-axis.

A graph is **symmetric with respect to the origin** if, whenever the point (x, y) is on the graph, then so is

_____ . To test an equation for this type of symmetry, replace _____ by

_____ and replace _____ by _____ . If the equation is

unchanged, then the graph is symmetric with respect to the origin.

Homework Assignment

Page(s)

Exercises

Name _____ Date _____

1.9   Graphing Calculators; Solving Equations and Inequalities Graphically

I. Using a Graphing Calculator

A graphing calculator's **viewing rectangle** is _____

_____.

Why is it important to choose a viewing rectangle with care?

Describe what is meant by the **[a, b]** by **[c, d]** **viewing rectangle.**

If you want an overview of the essential features of a graph, you must choose a _____

_____ viewing rectangle to obtain a global view of the graph. If you want to investigate the

details of a graph, you must _____ viewing rectangle that shows just

the feature of interest.

II. Solving Equations Graphically

Describe how to solve an equation using the **graphical method.**

Describe advantages and disadvantages of solving equations by the algebraic method.

Describe advantages and disadvantages of solving equations by the graphical method.

III. Solving Inequalities Graphically

Describe how to solve the inequality $y \geq -x^2 + 4$ graphically.

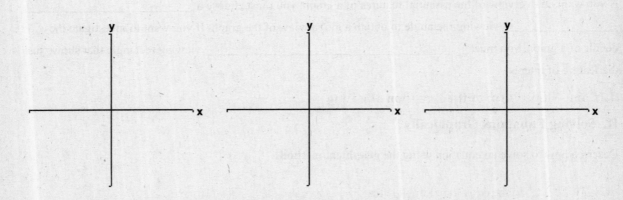

Homework Assignment

Page(s)

Exercises

Name _____ Date _____

1.10 Lines

I. The Slope of a Line

The "steepness" of a line refers to how quickly it _____

_____. We define *run* to be _____ and

we define *rise* to be _____.

The *slope* of a line is _____.

If a line lies in a coordinate plane, then the **run** is _____ and the

rise is _____ between any two points on the line.

The **slope** *m* of a nonvertical line that passes through the points $A(x_1, y_1)$ and $B(x_2, y_2)$ is

_____.

The slope of a vertical line is _____. The slope of a horizontal line is _____.

Lines with positive slope slant _____. Lines with negative slope

slant _____. The steepest lines are those for which _____

_____.

Example 1: Find the slope of the line that passes through the points (1, 9) and (15, 2).

II. Point-Slope Form of the Equation of a Line

The **point-slope form** of the equation of the line that passes through the point (x_1, y_1) and has slope *m* is

_____ .

Example 2: Find the equation of the line that passes through the points (1, 9) and (15, 2).

III. Slope-Intercept Form of the Equation of a Line

The **slope-intercept form** of the equation of a line that has slope m and y-intercept b is

_____.

Example 3:   Find the equation of the line with slope -5 and y-intercept -3.

IV. Vertical and Horizontal Lines

The equation of the vertical line through (a, b) is _____. The equation of the horizontal line through (a, b) is _____.

V. General Equation of a Line

The general equation of a line is given as _____, where A and B are _____ _____. The graph of every linear equation $Ax + By + C = 0$ is a _____. Conversely, every line is the graph of _____.

VI. Parallel and Perpendicular Lines

Two nonvertical lines are parallel if and only if _____.

Two lines with slopes m_1 and m_2 are perpendicular if and only if _____, that is, their

slopes are _____: $m_2 = -\dfrac{1}{m_1}$. Also, a horizontal line, having slope

0, is perpendicular to a _____.

Example 4:   Find the equation of the line that is parallel to the line $2x - y = 4$ and passes through the point $(0, 5)$.

VII. Modeling with Linear Equations: Slope as Rate of Change

When a line is used to model the relationship between two quantities, the slope of the line is the _____

_____ of one quantity with respect to the other.

Give an example of a real-life situation in which the slope of a line is a rate of change.

Additional notes

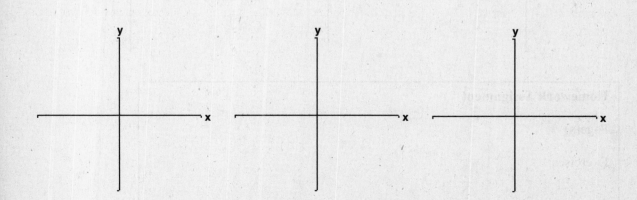

Additional notes

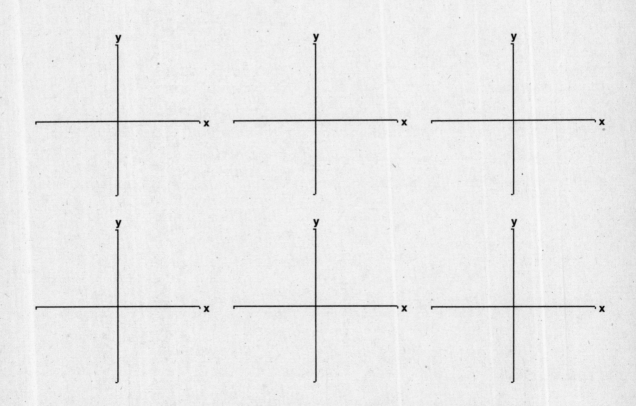

Homework Assignment

Page(s)

Exercises

Name _____ Date _____

1.11 Making Models Using Variation

I. Direct Variation

If the quantities x and y are related by an equation _____ for some constant

$k \neq 0$, we say that y **varies directly as x**, or _____, or

simply _____. The constant k is called the _____

_____.

Example 1: Suppose that w is directly proportional to t. If w is 21 when t is 6, what is the value of w when t is 19?

II. Inverse Variation

If the quantities x and y are related by the equation $y = \dfrac{k}{x}$ for some constant $k \neq 0$, we say that y is _____

_____ or _____.

Example 2: Suppose that w is inversely proportional to t. If w is 21 when t is 6, what is the value of w when t is 9?

III. Joint Variation

If one quantity is proportional to two or more other quantities, we call this relationship

_____.

If the quantities x, y, and z are related by the equation _____ where k is a

nonzero constant, we say that z **varies jointly as** x and y or_____

_____.

Example 3:   Suppose that z is jointly proportional to x and y. If z is 45 when $x = -3$ and $y = 5$, what is the value of z when $x = 6$ and $y = \dfrac{1}{2}$?

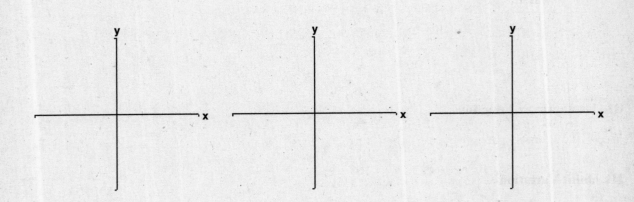

Homework Assignment

Page(s)

Exercises

Chapter 2   Functions

2.1   What Is a Function?

I. Functions All Around Us

Give a real-life example of a function.

II. Definition of Function

A **function** f is a _____

_____.

The symbol $f(x)$ is read _____ and is called the _____,

or the _____.

The set A is called the _____ of the function, and the **range** of f is _____

_____.

The symbol that represents an arbitrary number in the domain of a function f is called _____

_____. The symbol that represents a number in the range of f is called _____

_____. If we write $f(x) = y$, then _____ is the independent variable and _____ is

the dependent variable.

III. Evaluating a Function

To evaluate a function f at a number, _____.

Example 1:   If $f(x) = 50 - 2x^2$, then evaluate $f(5)$.

A **piecewise-defined function** is _____
_____.

IV. The Domain of a Function

The domain of a function is _____. If the
function is given by an algebraic expression and the domain is not stated explicitly, then by convention the
domain of the function is _____
_____.

Example 2:   Find the domain of the function $g(x) = \sqrt{x^2 - 16}$

V. Four Ways to Represent a Function

List and describe the four ways in which a specific function can be described.

Homework Assignment

Page(s)

Exercises

Name _____ Date _____

2.2   Graphs of Functions

I. Graphing Functions by Plotting Points

To graph a function f, _____

_____

_____ .

If f is a function with domain A, then the **graph** of f is the set of ordered pairs _____,

plotted in a coordinate plane. In other words, the graph of f is the set of all points (x, y) such that _____;

that is, the graph of f is the graph of the equation _____ .

A function f of the form $f(x) = mx + b$ is called a _____ because its graph is

the graph of the equation $y = mx + b$, which represents a line with slope _____ and y-intercept

_____ . The function $f(x) = b$, where b is a given number, is called a _____

_____ because all its values are _____ .

Its graph is _____ .

II. Graphing Functions with a Graphing Calculator

Describe how to use a graphing calculator to graph the function $f(x) = 5x^3 - 2x + 2$.

III. Graphing Piecewise Defined Functions

Describe how to graph the piecewise-defined function $f(x) = \begin{cases} 3 - \dfrac{1}{5}x, & \text{if } x < 0 \\ 2x^2, & \text{if } x \geq 0 \end{cases}$.

The **greatest integer function** is defined by $[\![\, x \,]\!] = $ _____.

The greatest integer function is an example of a _____.

A function is called **continuous** if _____.

IV. The Vertical Line Test

The **Vertical Line Test** states that _____
_____.

Is the graph below the graph of a function? Explain.

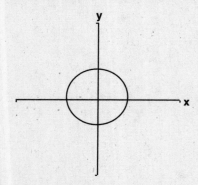

V. Equations That Define Functions

Any equation in the variables x and y defines a relationship between these variables. Does every equation in x and y define y as a function of x? _____

Draw an example of the graph of each type of function.

Linear Function

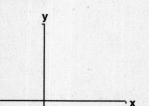

Power Function

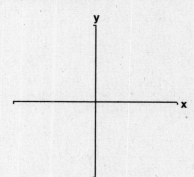

Root Function

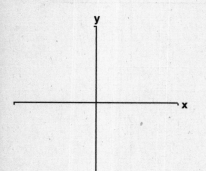

Reciprocal Function

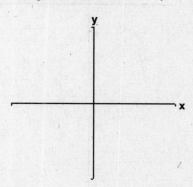

Absolute Value Function

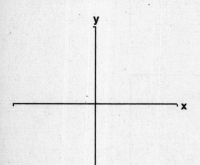

Greatest Integer Function

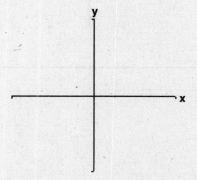

Additional notes

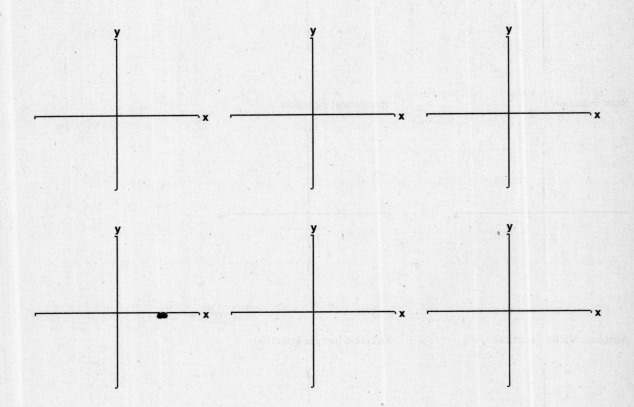

Homework Assignment

Page(s)

Exercises

Name _____ Date _____

2.3 Getting Information from the Graph of a Function

I. Values of a Function; Domain and Range

To analyze the graph of a function, _____
_____.

Describe how to use the graph of a function to find the function's domain and range.

II. Increasing and Decreasing Functions

A function *f* is said to be *increasing* when _____ and is said to be

decreasing when _____.

According to the definition of increasing and decreasing functions, *f* is **increasing** on an interval *I* if

_____ whenever _____ in *I*. Similarly, *f* is **decreasing**

on an interval *I* if _____ whenever _____ in *I*.

Example 1: Use the graph to determine (a) the domain, (b) the range, (c) the intervals on which the function is increasing, and (d) the intervals on which the function is decreasing.

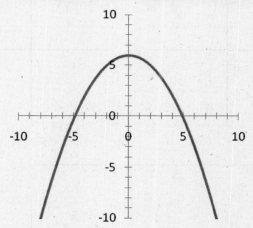

III. Local Maximum and Minimum Values of a Function

The function value $f(a)$ is a **local maximum value** of f if _____ when x is near a. In this case we say that f has _____.

The function value $f(a)$ is a **local minimum value** of f if _____ when x is near a. In this case we say that f has _____.

Describe how to use a graphing calculator to find the local maximum and minimum values of a function.

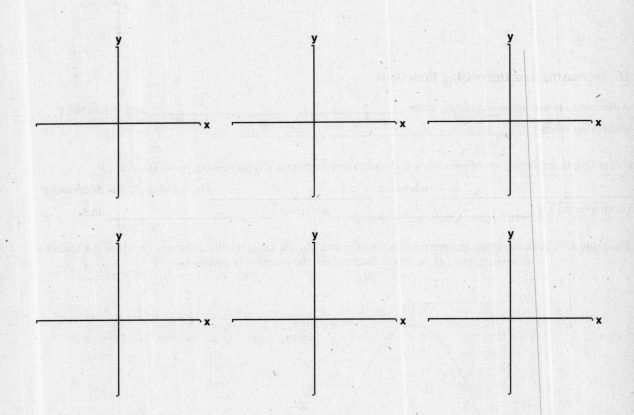

Homework Assignment

Page(s)

Exercises

Name _____ Date _____

2.4 Average Rate of Change of a Function

I. Average Rate of Change

The **average rate of change** of the function $y = f(x)$ between $x = a$ and $x = b$ is

_____

The average rate of change is the slope of the _____ between $x = a$ and $x = b$ on the
graph of f, that is, the line that passes through _____.

Example 1: For the function $f(x) = 3x^2 - 2$, find the average rate of change between $x = 2$ and $x = 4$.

II. Linear Functions Have Constant Rate of Change

For a linear function $f(x) = mx + b$, the average rate of change between any two points is _____

_____. If a function f has constant average rate of change, then it must be _____

_____.

Example 2: For the function $f(x) = 14 - 6x$, find the average rate of change between the following points.
  (a) $x = -10$ and $x = -5$
  (b) $x = 0$ and $x = 3$
  (c) $x = 4$ and $x = 9$

Additional notes

Homework Assignment

Page(s)

Exercises

Name _____ Date _____

2.5   Transformations of Functions

I. Vertical Shifting

Adding a constant to a function shifts its graph _____: upward if the constant is

_____ and downward if it is _____.

Consider vertical shifts of graphs. Suppose $c > 0$. To graph $y = f(x) + c$, shift _____

_____. To graph $y = f(x) - c$, shift _____

_____.

II. Horizontal Shifting

Consider horizontal shifts of graphs. Suppose $c > 0$. To graph $y = f(x - c)$, shift _____

_____. To graph $y = f(x + c)$, shift _____

_____.

III. Reflecting Graphs

To graph $y = -f(x)$, reflect the graph of $y = f(x)$ in the _____.

To graph $y = f(-x)$, reflect the graph of $y = f(x)$ in the _____.

IV. Vertical Stretching and Shrinking

Multiplying the y-coordinates of the graph of $y = f(x)$ by c has the effect of _____

_____.

To graph $y = cf(x)$:

If $c > 1$, _____

If $0 < c < 1$, _____.

V. Horizontal Stretching and Shrinking

To graph $y = f(cx)$:

If $c > 1$, _____.

If $0 < c < 1$, _____.

VI. Even and Odd Functions

Let f be a function.

Then f is **even** if _____.

Then f is **odd** if _____.

The graph of an even function is symmetric with respect to _____.

The graph of an odd function is symmetric with respect to _____.

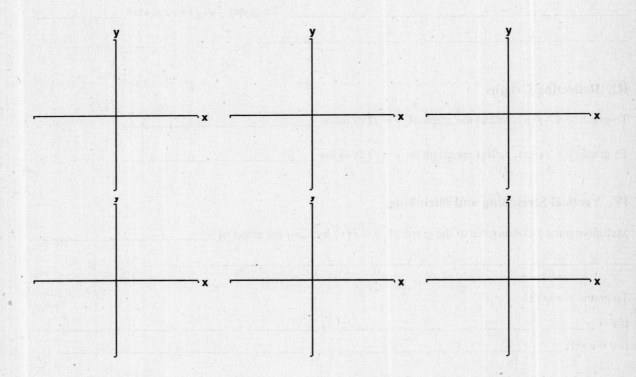

Homework Assignment

Page(s)

Exercises

Name _____ Date _____

2.6 Combining Functions

I. Sums, Differences, Products, and Quotients

Given two functions f and g, we define the new function $f + g$ by _____.

The new function $f + g$ is called _____. Its value at

x is _____.

Let f and g be functions with domains A and B. Then the functions $f + g$, $f - g$, fg, and f/g are defined as follows.

$(f + g)(x) = $ _____, Domain is _____

$(f - g)(x) = $ _____, Domain is _____

$(fg)(x) = $ _____, Domain is _____

$\left(\dfrac{f}{g}\right)(x) = $ _____, Domain is _____

The graph of the function $f + g$ can be obtained from the graphs of f and g by **graphical addition,** meaning that

we _____.

Example 2:   Let $f(x) = 3x + 1$ and $g(x) = 2x^2 - 1$.
   (a) Find the function $g - f$.
   (b) Find the function $f + g$.

II. Composition of Functions

Given two functions f and g, the **composite function** $f \circ g$ (also called _____

_____) is defined by _____ .

The domain of $f \circ g$ is _____

_____ . In other words, $(f \circ g)(x)$ is defined whenever _____

_____ .

Example 2:   Let $f(x) = 3x + 1$ and $g(x) = 2x^2 - 1$.
(a) Find the function $f \circ g$.
(b) Find $(f \circ g)(2)$.

It is possible to take the composition of three or more functions. For instance, the composite function $f \circ g \circ h$

is found by _____ .

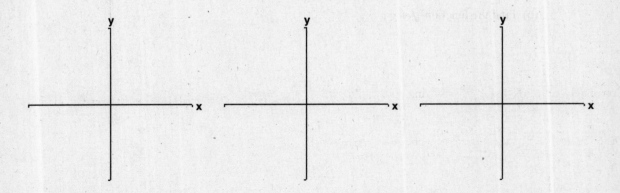

Homework Assignment

Page(s)

Exercises

Name _____ Date _____

2.7 One-to-One Functions and Their Inverses

I. One-to-One Functions

A function with domain A is called a **one-to-one function** if _____

_____.

An equivalent way of writing the condition for a one-to-one function is this:

_____.

The **Horizontal Line Test** states that _____

_____.

Every increasing function and every decreasing function is _____.

II. The Inverse of a Function

Let f be a one-to-one function with domain A and range B. Then its _____ f^{-1}

has domain B and range A and is defined by $f^{-1}(y) = x \iff f(x) = y$ for any y in B.

Let f be a one-to-one function with domain A and range B. The inverse function f^{-1} satisfies the following cancellation properties:

    1)

    2)

Conversely, any function f^{-1} satisfying these equtions is _____.

Describe how to find the inverse of a one-to-one function.

Example 1:   Find the inverse of the function $f(x) = 9 - 2x$

III. Graphing the Inverse of a Function

The graph of f^{-1} is obtained by _____.

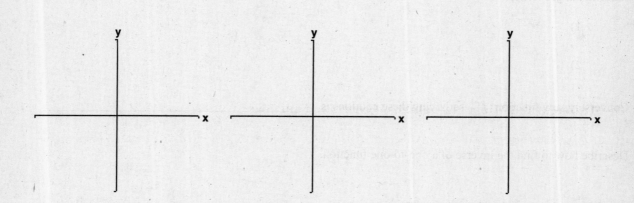

Homework Assignment

Page(s)

Exercises

Chapter 3   Polynomial and Rational Functions

3.1   Quadratic Functions and Models

A **polynomial function of degree *n*** is a function of the form _____.

A **quadratic function** is a polynomial function of degree _____. A quadratic function is a function of the form _____.

I. Graphing Quadratic Functions Using the Standard Form

A quadratic function $f(x) = ax^2 + bx + c$ can be expressed in the **standard form** $f(x) = a(x-h)^2 + k$ by

_____. The graph of *f* is a _____ with **vertex**

_____. The parabola opens upward if _____ or downward

if _____.

Example 1:   Let $f(x) = 3x^2 + 6x - 1$.
    (a) Express *f* in standard form.
    (b) What is the vertex of the graph of *f*?
    (c) Does the graph of *f* open upward or downward?

II. Maximum and Minimum Values of Quadratic Functions

If a quadratic function has vertex (h, k), then the function has a minimum value at the vertex if its graph opens

_____ and a maximum value at the vertex if its graph opens _____.

Let *f* be a quadratic function with standard form $f(x) = a(x-h)^2 + k$. The maximum or minimum value of *f*

occurs at _____.

If $a > 0$, then the **minimum value** of *f* is _____.

If $a < 0$, then the **maximum value** of *f* is _____.

Example 2:   Consider the quadratic function $f(x) = -2x^2 + 4x - 8$.
         (a) Express f in standard form.
         (b) Does f have a minimum value or a maximum value? Explain.
         (c) Find the minimum or maximum value of f.

The maximum or minimum value of a quadratic function $f(x) = ax^2 + bx + c$ occurs at _____.

If $a > 0$, then the _____ value is $f\left(-\dfrac{b}{2a}\right)$.

If $a < 0$, then the _____ value is $f\left(-\dfrac{b}{2a}\right)$.

Example 3:   Find the maximum or minimum value of the quadratic function $f(x) = 0.5x^2 - 5x + 12$, and state whether it is the maximum or the minimum.

Homework Assignment

Page(s)

Exercises

Name _____ Date _____

3.2   Polynomial Functions and Their Graphs

A **polynomial function of degree n** is a function of the form _____

where n is a nonnegative integer and $a_n \neq 0$.

The numbers $a_0, a_1, a_2, \ldots, a_n$ are called the _____ of the polynomial.

The number a_0 is the _____ or _____.

The number a_n, the coefficient of the highest power, is the _____, and

the term $a_n x^n$ is the _____.

I. Graphing Basic Polynomial Functions

The graphs of polynomials of degree 0 or 1 are _____, and the graphs of polynomials of degree

2 are _____. The greater the degree of a polynomial, the more _____

_____. However, the graph of a polynomial function is _____,

meaning that the graph has no _____. In addition, the graph of a polynomial function

is a _____ curve; that is, it has no _____.

The graph of $P(x) = x^n$ has the same general shape as the graph of _____ when n is even and

the same general shape as the graph of $y = x^3$ when _____.

II. End Behavior and the Leading Term

The **end behavior** of a polynomial is _____

_____. To describe end behavior we use $x \to \infty$ to

mean _____ and we use $x \to -\infty$ to mean _____

_____.

The end behavior of the polynomial $P(x) = a_n x^n + a_{n-1} x^{n-1} + \cdots + a_1 x + a_0$ is determined by _____

_____.

In your own words, describe the end behavior of a polynomial P with the following characteristics:

If P has odd degree and a positive leading coefficient, then _____

_____.

If P has odd degree and a negative leading coefficient, then _____

_____.

If P has even degree and a positive leading coefficient, then _____

_____.

If P has even degree and a negative leading coefficient, then _____

_____.

Example 1:   Determine the end behavior of the polynomial $P(x) = -5x^6 - 2x^4 + 3x^2 - 9$.

III. Using Zeros to Graph Polynomials

If P is a polynomial function, then c is called a _____ of P if $P(c) = 0$. In other words, the

zeros of P are the solutions of _____. Note that if

$P(c) = 0$, then the graph of P has an x-intercept at _____, so the x-intercepts of the graph are

the zeros of the function.

If P is a polynomial and c is a real number, then list four equivalent statements about the real zeros of P.

The **Intermediate Value Theorem for Polynomials** states that _____

_____

_____.

List guidelines for graphing polynomial functions.

IV. Shape of the Graph Near a Zero

If c is a zero of P, and the corresponding factor $x - c$ occurs exactly m times in the factorization of P then we say that c is a _____.

The graph crosses the x-axis at c if the multiplicity m is _____ and does not cross the x-axis if m is _____.

Example 2: How many times does the graph of $P(x) = (x-3)^2(x+5)(x-9)^3$ cross the x-axis?

V. Local Maxima and Minima of Polynomials

If the point $(a, f(a))$ is the highest point on the graph of f within some viewing rectangle, then $f(a)$ is a local maximum value of f and the point $(a, f(a))$ is a _____ on the graph. If the point $(b, f(b))$ is the lowest point on the graph of f within some viewing rectangle, then $f(b)$ is a local minimum value of f and the point $(b, f(b))$ is a _____ on the graph.

The set of all local maximum and minimum points on the graph of a function is called its

_____. If $P(x) = a_n x^n + a_{n-1} x^{n-1} + \cdots + a_1 x + a_0$ is a polynomial of degree n, then the graph of P has at most _____ local extrema.

Example 3: How many local extrema does the polynomial $P(x) = -x^4 - x^3 + x^2 - 4x + 2$ have?

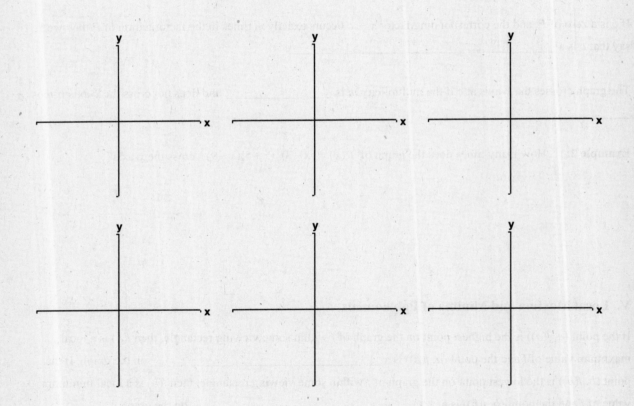

Name _____ Date _____

3.3 Dividing Polynomials

I. Long Division of Polynomials

Describe the Division Algorithm.

The division process ends when _____.

Example 1: Show how to set up the long division of $3x^3 - 1$ by $2x + 5$.

II. Synthetic Division

Synthetic division is _____. It can be
used when the divisor is of the form _____.

Example 2: Show how to set up the synthetic division of $4x^3 + 7x - 21$ by $x + 2$.

III. The Remainder and Factor Theorems

The **Remainder Theorem** states that if the polynomial $P(x)$ is divided by $x - c$, then the remainder is the
value _____.

The **Factor Theorem** states that c is a zero of P if and only if _____.

Explain how to easily decide whether $x - 6$ is a factor of the polynomial $P(x) = x^7 - 5x^5 + 2x^4 - x^2 + 9$ without performing long division.

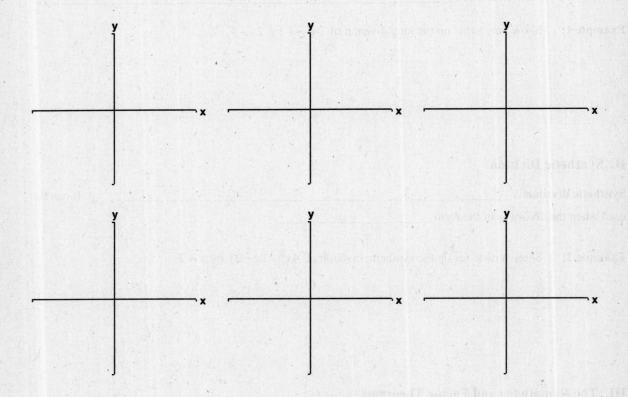

Homework Assignment

Page(s)

Exercises

Name _____ Date _____

3.4 Real Zeros of Polynomials

I. Rational Zeros of Polynomials

State the Rational Zeros Theorem.

We see from the Rational Zeros Theorem that if the leading coefficient is 1 or -1, then the rational zeros must

be _____.

List the steps for finding the rational zeros of a polynomial.

II. Descartes' Rule of Signs and Upper and Lower Bounds for Roots

If $P(x)$ is a polynomial with real coefficients, written with descending powers of x (and omitting powers with

coefficient 0), then a **variation in sign** occurs whenever _____

_____.

State Descartes' Rule of Signs.

We say that a is a **lower bound** and b is an **upper bound** for the zeros of a polynomial if _____
_____.

State the Upper and Lower Bounds Theorem.

The phrase "alternately nonpositive and nonnegative" simply means _____
_____.

III. Using Algebra and Graphing Devices to Solve Polynomial Equations

Describe how to use the algebraic techniques from this section to select an appropriate viewing rectangle when solving a polynomial equation graphically.


```
Homework Assignment

Page(s)

Exercises
```

Name _____ Date _____

3.5 Complex Numbers

To make it possible to solve *all* quadratic equations, mathematicians invented an expanded number system,

called the _____. They defined the number $i =$ _____,

so that $i^2 =$ _____.

A **complex number** is an expression of the form _____ where a and b are real numbers

and $i^2 =$ _____. The **real part** of this complex number is _____ and the **imaginary**

part is _____. Two complex numbers are **equal** if and only if _____

_____.

A complex number in which the real part is 0 is called a _____. A real

number such as -3 can be thought of as a complex number with _____.

I. Arithmetic Operations on Complex Numbers

Describe how to add complex numbers.

Describe how to subtract complex numbers.

Describe how to multiply complex numbers.

$(a+bi)+(c+di) =$ _____

$(a+bi)-(c+di) =$ _____

$(a+bi)\cdot(c+di) =$ _____

Division of complex numbers is much like _____

_____.

For the complex number $z = a+bi$, we define its **complex conjugate** to be $\overline{z} =$ _____.

$z \cdot \overline{z} = (a+bi)(a-bi) = $ _____

Notice that the product of a complex number and its conjugate is always _____

_____ .

Dividing Complex Numbers

To simplify the quotient $\dfrac{a+bi}{c+di}$, multiply the numerator and the denominator by _____

_____ :

$$\frac{a+bi}{c+di} = \left(\frac{a+bi}{c+di}\right)\left(\underline{\hspace{2cm}}\right) = $$

_____   _____

II. Square Roots of Negative Numbers

If $-r$ is negative, then the **principal square root** of $-r$ is $\sqrt{-r} = $ _____

The two square roots of $-r$ are _____ and _____

When multiplying radicals of negative numbers, take care to express them first _____

_____ .

III. Complex Solutions of Quadratic Equations

For $ax^2 + bx + c = 0$ and $a \neq 0$, if $b^2 - 4ac < 0$, then the equation has _____ solution(s).

However, in the complex number system, this equation will *always* have solutions because _____

_____ .

If a quadratic equation with real coefficients has complex solutions, then these solutions are _____

_____ of each other. So if $a + bi$ is a solution then _____ is also a

solution.

Homework Assignment

Page(s)

Exercises

Name _____ Date _____

3.6   Complex Zeros and the Fundamental Theorem of Algebra

I. The Fundamental Theorem of Algebra

State the Fundamental Theorem of Algebra.

State the Complete Factorization Theorem.

To actually find the complex zeros of an nth-degree polynomial, we usually _____

_____.

Example 1:   Suppose the zeros of the fourth-degree polynomial P are 5, -2, $14i$, and $-14i$. Write the
complete factorization of P.

II. Zeros and Their Multiplicities

If the factor $x-c$ appears k times in the complete factorization of $P(x)$, then we say that c is a zero of

_____.

The Zeros Theorem states that every polynomial of degree $n \geq 1$ has exactly _____ zeros, provided that
a zero of multiplicity k is counted _____.

Example 2:   Suppose the zeros of P are 5 with multiplicity 1, -2 with multiplicity 3, $14i$ with multiplicity 2,
and $-14i$ with multiplicity 2. Write the complete factorization of P.

III. Complex Zeros Come in Conjugate Pairs

The Conjugate Zeros Theorem states that if the polynomial P has real coefficients and if the complex number z is a zero of P, then _____.

IV. Linear and Quadratic Factors

A quadratic polynomial with no real zeros is called _____.

Every polynomial with real coefficients can be factored into _____

_____.

Homework Assignment

Page(s)

Exercises

Name _____ Date _____

3.7 Rational Functions

I. Rational Functions and Asymptotes

A rational function is a function of the form _____ where $P(x)$ and $Q(x)$ are polynomial

functions having no factors in common.

The domain of a rational function consists of _____

_____. When graphing a rational function, we must pay special attention

to the behavior of the graph near _____.

$x \to a^-$ means _____

$x \to a^+$ means _____

$x \to -\infty$ means _____

$x \to \infty$ means _____

Informally speaking, an asymptote of a function is _____

_____.

The line $x = a$ is a **vertical asymptote** of the function $y = f(x)$ if _____

_____.

Draw an example of a graph having a vertical asymptote.

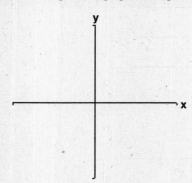

The line $y = b$ is a **horizontal asymptote** of the function $y = f(x)$ if _____

_____.

Draw an example of a graph having a horizontal asymptote.

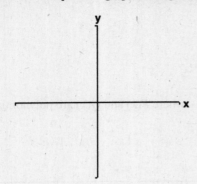

A rational function has vertical asymptotes where _____

_____ .

II. Transformations of $y = 1/x$

A rational function of the form $r(x) = \dfrac{ax+b}{cx+d}$ can be graphed by shifting, stretching, and/or reflecting _____

_____ .

III. Asymptotes of Rational Functions

Let r be the rational function $r(x) = \dfrac{a_n x^n + a_{n-1} x^{n-1} + \cdots + a_1 x + a_0}{b_m x^m + b_{m-1} x^{m-1} + \cdots + b_1 x + b_0}$.

1. The vertical asymptotes of r are the lines _____.

2. (a) If $n < m$, then r has _____.

    (b) If $n = m$, then r has _____.

    (c) If $n > m$, then r has _____.

IV. Graphing Rational Functions

List the guidelines for sketching graphs of rational functions.

When graphing a rational function, describe how to check for the graph's behavior near a vertical asymptote.

Is it possible for a graph to cross a vertical asymptote?

Is it possible for a graph to cross a horizontal asymptote?

V. Slant Asymptotes and End Behavior

If $r(x) = P(x)/Q(x)$ is a rational function in which the degree of the numerator is _____

_____, we can use the Division Algorithm to express the function in the

form $r(x) = ax + b + \dfrac{R(x)}{Q(x)}$, where the degree of R is less than the degree of Q and $a \neq 0$. For large values of

$|x|$, the graph of $y = r(x)$ approaches the graph of _____. In this situation

we say that $y = ax + b$ is a _____.

Describe how to graph a rational function which has a slant asymptote.

Additional notes

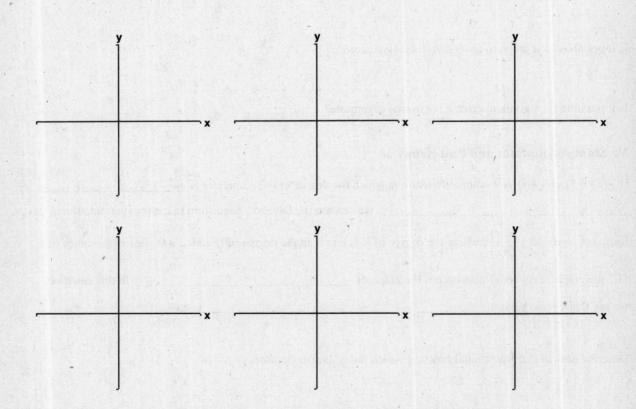

Homework Assignment

Page(s)

Exercises

Chapter 4   Exponential and Logarithmic Functions

4.1   Exponential Functions

I. Exponential Functions

True or false? The Laws of Exponents are true when the exponents are real numbers.

The **exponential function with base a** is defined for all real numbers x by _____,

where $a > 0$ and $a \neq 1$.

Example 1:   Let $f(x) = 6^x$. Evaluate the following:
        (a) $f(3)$
        (b) $f(-2)$
        (c) $f(\sqrt{3})$

II. Graphs of Exponential Functions

The exponential function $f(x) = a^x$, $(a > 0, a \neq 1)$ has domain _____ and range _____.

The line $y = 0$ (the x-axis) is a _____. If $0 < a < 1$, then f

_____ rapidly. If $a > 1$, then f _____ rapidly.

Example 2:   Sketch the graph of the function $f(x) = 3^{-1}$.

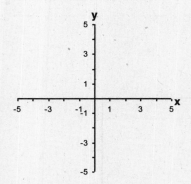

III. Compound Interest

In terms of an investment earning compound interest, the _____ P is the amount of money that is initially invested.

Compound interest is calculated by the formula _____

where $A(t) =$ _____

              $P =$ _____

              $r =$ _____

              $n =$ _____

              $t =$ _____

If an investment earns compound interest, then the **annual percentage yield** (APY) is the _____

_____.

Homework Assignment

Page(s)

Exercises

Name _____ Date _____

4.2 The Natural Exponential Function

I. The Number e

The number e is defined as the value that $\left(1+\dfrac{1}{n}\right)^n$ _____.

The approximate value of e is _____.

In can be shown that e is a(n) _____, so we cannot write its exact value in decimal form.

II. The Natural Exponential Function

The **natural exponential function** is the exponential function _____ with base e. It is often referred to as *the* exponential function.

Example 1: Evaluate the expression correct to five decimal places: $4e^{0.25}$

III. Continuously Compounded Interest

Continuously compounded interest is calculated by the formula _____

where $\quad A(t) =$ _____

$\qquad\qquad P =$ _____

$\qquad\qquad r =$ _____

$\qquad\qquad t =$ _____

Example 2:   Find the amount after 10 years if $5000 is invested at an interest rate of 8% per year, compounded continuously.

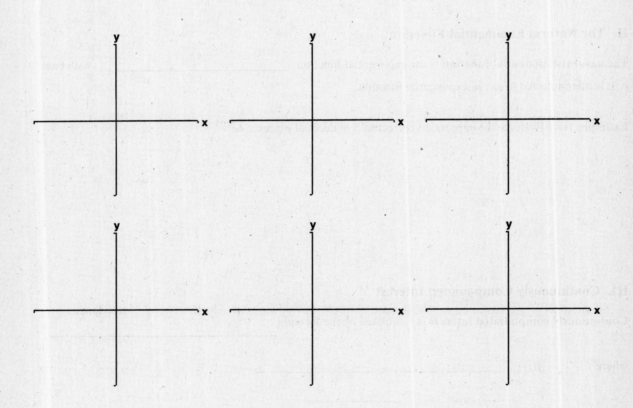

Homework Assignment

Page(s)

Exercises

Name _____ Date _____

4.3 Logarithmic Functions

I. Logarithmic Functions

Every exponential function $f(x) = a^x$, with $a > 0$ and $a \neq 1$, is a _____ by

the Horizontal Line Test and therefore has _____. The

inverse function f^{-1} is called the _____ and

is denoted by $\log_a$.

For the definition of the logarithmic function, let a be a positive number with $a \neq 1$. The **logarithmic function**

with base a, denoted by $\log_a$, is defined by

_____

So $\log_a x$ is the _____ to which the base a must be raised to give _____.

Complete each of the following properties of logarithms.

1. $\log_a 1 = $ _____

2. $\log_a a = $ _____

3. $\log_a a^x = $ _____

4. $a^{\log_a x} = $ _____

II. Graphs of Logarithmic Functions

Recall that if a one-to-one function f has domain A and range B, then its inverse function f^{-1} has domain

_____ and range _____. Since the exponential function $f(x) = a^x$ with $a \neq 1$ has

domain $\mathbb{R}$ and range $(0, \infty)$, we conclude that its inverse function, $f^{-1}(x) = \log_a x$, has domain

_____ and range _____.

The graph of $f^{-1}(x) = \log_a x$ is obtained by _____

_____.

The x-intercept of the function $y = \log_a x$ is _____. The _____ is

a vertical asymptote of $y = \log_a x$.

Example 1:   Sketch the graph of the function $f(x) = \log_3 x$.

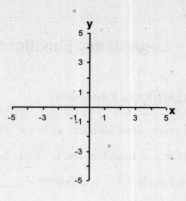

III. Common Logarithms

The logarithm with base 10 is called the _____ and is denoted by

_____.

Example 2:   Evaluate $\log 30$.

IV. Natural Logarithms

The logarithm with base e is called the _____ and is denoted by

_____.

The natural logarithmic function $y = \ln x$ is the inverse function of _____

_____.

Complete each of the following properties of natural logarithms.

1. $\ln 1 =$ _____

2. $\ln e =$ _____

3. $\ln e^x =$ _____

4. $e^{\ln x} =$ _____

Example 3:   Evaluate $\ln 30$.

Homework Assignment

Page(s)

Exercises

Name _____ Date _____

4.4 Laws of Logarithms

I. Laws of Logarithms

Let a be a positive number, with $a \neq 1$. Let A, B, and C be any real numbers with $A > 0$ and $B > 0$. Complete each of the following Laws of Logarithms and give a description of each law.

1. $\log_a(AB) = $ _____

2. $\log_a\left(\dfrac{A}{B}\right) = $ _____

3. $\log_a(A^c) = $ _____

II. Expanding and Combining Logarithmic Expressions

The process of writing the logarithm of a product or a quotient as the sum or different of logarithms is called

_____.

The process of writing sums and differences of logarithms as a single logarithms is called

_____.

Note that although the Laws of Logarithms tell us how to compute the logarithm of a product or quotient, there is no corresponding rule for the _____.

Example 1: Expand the logarithmic expression $\ln\dfrac{x^4 y}{3}$.

Example 2: Condense the logarithmic expression $2\log w + 3\log(2w+1)$.

III. Change of Base Formula

The Change of Base Formula is given by

_____

Describe the advantage to using the Change of Base Formula.

Explain how to use a calculator to evaluate $\log_{13} 150$.

Homework Assignment

Page(s)

Exercises

Name _____ Date _____

4.5 Exponential and Logarithmic Equations

I. Exponential Equations

An exponential equation is one in which _____.

List the guidelines for solving exponential equations.

Example 1: Find the solution of the equation $5^{2x-1} = 20$, correct to six decimal places.

II. Logarithmic Equations

A logarithmic equation is one in which _____.

List the guidelines for solving logarithmic equations.

Example 2: Solve $12\ln(x+5) - 2 = 22$ for x, correct to six decimal places.

III. Compound Interest

Describe how to find the length of time rquired for an investment to double.

Example 3:   How long will it take for a $12,000 investment to double if it is invested at an interest rate of 6% per year and if the interest is compounded continuously.

Homework Assignment

Page(s)

Exercises

Name _____ Date _____

4.6 Modeling with Exponential and Logarithmic Functions

I. Exponential Growth (Doubling Time)

If the initial size of a population is n_0 and the doubling time is a, then the size of the population at time t is given

by _____ where a and t are measured in the same time units (minutes, hours,

days, years, and so on).

Example 1:   Suppose a community's population doubles every 20 years. Initially there are 500 members in
the community.
(a) Find a model for the community's population after t years.
(b) How many community members are there after 30 years?
(c) When will the population of the community reach 10,000?

II. Exponential Growth (Relative Growth Rate)

A population's **relative growth rate** r is the _____

_____.

A population that experiences exponential growth increases according to the model _____ where

   $n(t) =$ _____,

   $n_0 =$ _____,

   $r =$ _____,

   $t =$ _____.

Example 2: The initial population of a colony is 800. If the colony has a relative growth rate of 15% per year, find a function that models the population after t years.

III. Radioactive Decay

The rate of radioactive decay is proportional to _____.

Physicists express the rate of radioactive decay in terms of _____.

The **radioactive decay model** states that if m_0 is the initial mass of a radioactive substance with half-life h, then the mass remaining at time t is modeled by the function _____, where

$r =$ _____.

IV. Newton's Law of Cooling

Newton's Law of Cooling states that the rate at which an object cools is _____

_____

_____.

Newton's Law of Cooling

If D_0 is the initial temperature difference between an object and its surroundings, and if its surroundings have temperature T_s, then the temperature of the object at time t is modeled by the function

_____

where k is a positive constant that depends on the type of object.

V. Logarithmic Scales

When a physical quantity varies over a very large range, it is often convenient to _____

_____ in order to have a more manageable set of numbers.

The acidity of a solution is given by its pH, defined as _____, where $[H^+]$ is the concentration of hydrogen ions measured in moles per liter (M).

The **Richter Scale** defines the magnitude M of an earthquake to be _____, where I is the _____ of the earthquake (measured by the amplitude of a seismograph reading taken 100 km from the epicenter of the earthquake) and S is the _____

_____ (whose amplitude is 1 micron = 10^{-4} cm).

The magnitude of a standard earthquake is _____.

According to the Decibel Scale, the **intensity level B**, measured in decibels (dB), is defined as

_____

where I_0 is a reference intensity and $I_0 = 10^{-12}$ W/m^2 (watts per square meter).

The intensity level of the barely audible reference sound is _____.

The threshold of pain is about _____.

Additional notes

Chapter 5   Trigonometric Functions: Unit Circle Approach

5.1   The Unit Circle

I. The Unit Circle

The unit circle is _____.

Its equation is _____.

Example 1:   Show that the point $P\left(\dfrac{\sqrt{2}}{2}, \dfrac{\sqrt{2}}{2}\right)$ is on the unit circle.

II. Terminal Points on the Unit Circle

Suppose t is a real number. Mark off a distance t along the unit circle, starting at the point $(1, 0)$ and moving in

a _____ if t is positive or in a _____

_____ if t is negtaive. In this way we arrive at a point $P(x, y)$, called the _____

_____ determined by the real number t.

The circumference of the unit circle is $C =$ _____. To move a point halfway around the unit

circle, it travels a distance of _____. To move a quarter of the distance around the circle,

it travels a distance of _____.

List the terminal point determined by each given value of t.

t	Terminal point determined by t
0	
$\dfrac{\pi}{6}$	
$\dfrac{\pi}{4}$	
$\dfrac{\pi}{3}$	
$\dfrac{\pi}{2}$	

III. The Reference Number

Let t be a real number. The **reference number** $\bar{t}$ associated with t is _____

_____.

To find the reference number $\bar{t}$, it is helpful to know _____

_____. If the terminal point lies in quadrants I or

IV, where x is positive, we find $\bar{t}$ by _____.

If it lies in quadrants II or III, where x is negative, we find $\bar{t}$ by _____

_____.

List the steps for finding the terminal point P determined by any value of t.

Since the circumference of the unit circle is 2π, the terminal point determined by t is the same as that

determined by _____. In general, we can add or subtract _____ any

number of times without changing the terminal point determined by t.

Homework Assignment

Page(s)

Exercises

Name _____ Date _____

5.2   Trigonometric Functions of Real Numbers

I. The Trigonometric Functions

Let t be any real number and let $P(x, y)$ be the terminal point on the unit circle determined by t. Complete the definitions of each trigonometric function.

$\sin t =$ _____ $\cos t =$ _____

$\tan t =$ _____ $\csc t =$ _____

$\sec t =$ _____ $\cot t =$ _____

Because the trigonometric functions can be defined in terms of the unit circle, they are sometimes called the

_____.

Complete the following table of special values of the trigonometric function.

t	$\sin t$	$\cos t$	$\tan t$	$\csc t$	$\sec t$	$\cot t$
0						
$\dfrac{\pi}{6}$						
$\dfrac{\pi}{4}$						
$\dfrac{\pi}{3}$						
$\dfrac{\pi}{2}$						

The domain of the sine function is _____.

The domain of the cosine function is _____.

The domain of the tangent function is _____.

The domain of the secant function is _____.

The domain of the cotangent function is _____.

The domain of the cosecant function is _____.

II. Values of the Trigonometric Functions

To compute other values of the trigonometric functions, we first _____.

The signs of the trigonometric functions depend on _____

_____.

Complete the table listing the signs of the trigonometric functions.

Quadrant	Positive Functions	Negative Functions
I		
II		
III		
IV		

Since the trigonometric functions are defined in terms of the coordinates of terminal points, we can use the

_____ to find values of the trigonometric functions.

Example 1:   Find the value of $\tan\left(\dfrac{7\pi}{4}\right)$.

Complete each reciprocal relation.

$\csc t = $ _____ $\sec t = $ _____ $\cot t = $ _____

The odd trigonometric functions are _____.
The even trigonometric functions are _____.

III. Fundamental Identities

Complete each of the following trigonometric identities.

$\dfrac{1}{\sin t} = $ _____ $\dfrac{1}{\cos t} = $ _____ $\dfrac{1}{\tan t} = $ _____.

$\dfrac{\sin t}{\cos t} = $ _____ $\dfrac{\cos t}{\sin t} = $ _____.

$\sin^2 t + \cos^2 t = $ _____ $\tan^2 t + 1 = $ _____

$1 + \cot^2 t = $ _____

Homework Assignment

Page(s)

Exercises

Name _____ Date _____

5.3   Trigonometric Graphs

I. Graphs of Sine and Cosine

A function f is **periodic** if there is a positive number p such that _____.

The least such positive number (if it exists) is the _____ of f. If f has period p, then the graph of f on any interval of length p is called _____.

The functions sine and cosine have period _____.

Example 1:   Sketch the basic sine curve between 0 and 2π.

Example 2:   Sketch the basic cosine curve between 0 and 2π.

II. Graphs of Transformations of Sine and Cosine

For the functions $y = a\sin x$ and $y = a\cos x$, the number $|a|$ is called the _____ and is the largest value these functions attain.

The sine and cosine curves $y = a\sin kx$ and $y = a\cos kx$, $(k > 0)$, have amplitude $|a|$ and period _____. An appropriate interval on which to graph one complete period is

_____.

The value of k has the effect of _____ if $k > 1$ or the effect of

_____ if $k < 1$.

The sine and cosine curves $y = a \sin k(x - b)$ and $y = a \cos k(x - b)$, $(k > 0)$, have amplitude $|a|$, period $2\pi/k$,

and phase shift _____. An appropriate interval on which to graph one complete period is

_____.

Example 3:   Find the amplitude, period, and phase shift of $y = 2\cos\left(x - \dfrac{\pi}{6}\right)$:

III. Using Graphing Devices to Graph Trigonometric Functions

When using a graphing device to graph a function, it is important to _____

_____.

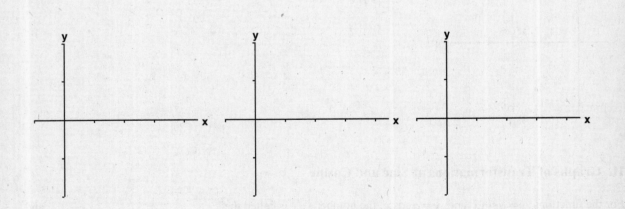

Homework Assignment

Page(s)

Exercises

Name _____ Date _____

5.4   More Trigonometric Graphs

I. Graphs of Tangent, Cotangent, Secant, and Cosecant

The functions tangent and cotangent have period _____.

The functions cosecant and secant have period _____.

The graph of $y = \tan x$ approaches the vertical lines $x = \dfrac{\pi}{2}$ and $x = -\dfrac{\pi}{2}$, so these lines are _____

_____.

The graph of $y = \cot x$ is undefined for _____, with n an integer, so its graph has

vertical asymptotes at these values.

To graph the cosecant and secant functions, we use the identities

_____

The graph of $y = \csc x$ has vertical asymptotes at _____.

The graph of $y = \sec x$ has vertical asymptotes at _____.

II. Graphs of Transformations of Tangent and Cotangent

The functions $y = a \tan kx$ and $y = a \cot kx$, $(k > 0)$, have period _____.

To graph one period of $y = a \tan kx$, an appropriate interval is _____.

To graph one period of $y = a \cot kx$, an appropriate interval is _____.

III. Graphs of Transformations of Cosecant and Secant

The functions $y = a \csc kx$ and $y = a \sec kx$, $(k > 0)$, have period _____.

An appropriate interval on which to graph one complete period is _____.

Additional notes

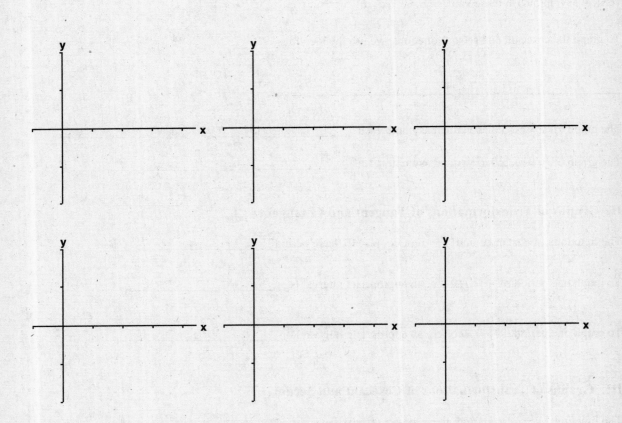

Homework Assignment

Page(s)

Exercises

Name _____ Date _____

5.5 Inverse Trigonometric Functions and Their Graphs

I. The Inverse Sine Function

The **inverse sine function** is the function $\sin^{-1}$ with domain _____ and range

_____ defined by $\sin^{-1} x = y \iff \sin y = x$.

The inverse sine function is also called _____, denoted by _____.

$y = \sin^{-1} x$ is the number in the interval $[-\pi/2, \pi/2]$ whose sine is _____.

The inverse sine function has these **cancellation properties**:

$\sin(\sin^{-1} x) = $ _____ for $-1 \le x \le 1$

$\sin^{-1}(\sin x) = $ _____ for $-\dfrac{\pi}{2} \le x \le \dfrac{\pi}{2}$

II. The Inverse Cosine Function

The **inverse cosine function** is the function $\cos^{-1}$ with domain _____ and

range _____ defined by $\cos^{-1} x = y \iff \cos y = x$.

The inverse cosine function is also called _____, denoted by _____.

$y = \cos^{-1} x$ is the number in the interval $[0, \pi]$ whose cosine is _____.

The inverse cosine function has these **cancellation properties**:

$\cos(\cos^{-1} x) = $ _____ for $-1 \le x \le 1$

$\cos^{-1}(\cos x) = $ _____ for $0 \le x \le \pi$

III. The Inverse Tangent Function

The **inverse tangent function** is the function $\tan^{-1}$ with domain _____ and range

_____ defined by $\tan^{-1} x = y \iff \tan y = x$.

The inverse tangent function is also called _____, denoted by _____.

$y = \tan^{-1} x$ is the number in the interval $\left(-\dfrac{\pi}{2}, \dfrac{\pi}{2} \right)$ whose tangent is _____.

The inverse tangent function has these **cancellation properties:**

$$\tan(\tan^{-1} x) = \underline{\hspace{3cm}} \text{ for } x \in \mathbb{R}$$

$$\tan^{-1}(\tan x) = \underline{\hspace{3cm}} \text{ for } -\frac{\pi}{2} < x < \frac{\pi}{2}$$

IV. The Inverse Secant, Cosecant, and Cotangent Functions

To define the inverse functions of the secant, cosecant, and cotangent functions, we restrict the domain of each function to _____

_____.

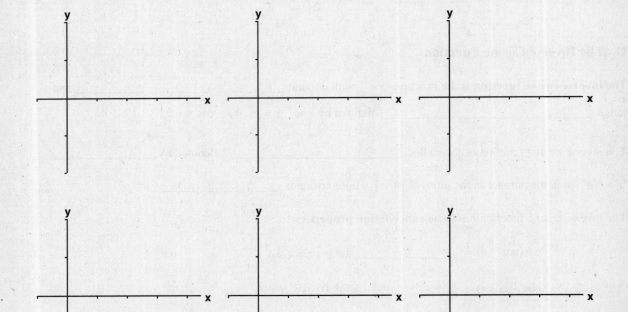

```
Homework Assignment

Page(s)

Exercises
```

Name _____ Date _____

5.6   Modeling Harmonic Motion

Periodic behavior is _____.

I. Simple Harmonic Motion

A **cycle** is _____.

If the equation describing the displacement y of an object at time t is $y = a \sin \omega t$ or $y = a \cos \omega t$, then the object is in _____.

In this case, the **amplitude,** which is the maximum displacement of the object, is given by _____.
The **period,** which is _____, is given by $\frac{2\pi}{\omega}$. The **frequency,** which is _____, is given by $\frac{\omega}{2\pi}$.

The functions $y = a \sin 2\pi vt$ or $y = a \cos 2\pi vt$ have frequency _____.

In general, the sine or cosine functions representing harmonic motion may be shifted horizontally or vertically. In this case, the equations take the form _____. The vertical shift b indicates that _____. The horizontal shift c indicates _____.

II. Damped Harmonic Motion

In a hypothetical frictionless environment, a spring will oscillate in such a way that its amplitude will not change. In the presence of friction, however, the motion of the spring eventually _____, that is, the amplitude of the motion _____. Motion of this type is called _____.

If the equation describing the displacement y of an object at time t is _____ or _____ $(c > 0)$, then the object is in **damped harmonic motion.** The constant c is the _____, _____ is the initial amplitude, and $2\pi / \omega$ is the _____.

Damped harmonic motion is simply harmonic motion for which the amplitude is governed by the function

_____.

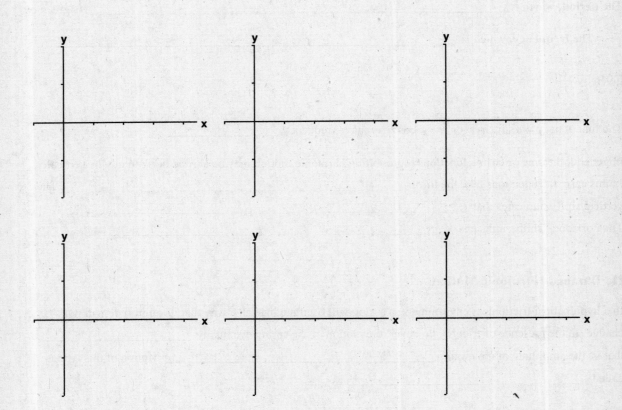

Homework Assignment

Page(s)

Exercises

Name _____ Date _____

Chapter 6   Trigonometric Functions: Right Triangle Approach

6.1   Angle Measure

An **angle** consists of two rays with a common _____.

We often interpret an angle as a rotation of the ray R_1 onto R_2. In this case, R_1 is called the _____

_____, and R_2 is called the _____ of the angle. If the

rotation is _____, the angle is considered **positive,** and if the rotation is

clockwise, the angle is considered _____.

I. Angle Measure

The **measure** of an angle is _____

_____.

One unit of measurement for angles is the **degree.** An angle of measure 1 degree is formed by _____

_____.

If a circle of radius 1 is drawn with the vertex of an angle at its center, then the measure of this angle in **radians**

(abbreviated rad) is _____.

For a circle of radius 1, a complete revolution has measure _____ rad; a straight angle has

measure _____ rad; and a right angle has measure _____ rad.

To convert degrees to radians, multiply by _____.

To convert radians to degrees, multiply by _____.

II. Angles in Standard Position

An angle is in **standard position** if it is drawn in the xy-plane with _____

_____.

Two angles in standard position are **coterminal** if _____.

III. Length of a Circular Arc

In a circle of radius r, the length s of an arc that subtends a central angle of θ radians is _____,

or, solving for θ, we get _____.

IV. Area of a Circular Sector

In a circle of radius r, the area A of a sector with a central angle of θ radians is _____.

V. Circular Motion

Linear speed is _____

_____.

Angular speed is _____

_____.

Suppose a point moves along a circle of radius r and the ray from the center of the circle to the point traverses θ radians in time t. Let $s = r\theta$ be the distance the point travels in time t. Then the speed of the object is given by

Angular speed _____   **Linear speed** _____

If a point moves along a circle of radius r with angular speed ω, then its linear speed v is given by

_____.

Homework Assignment

Page(s)

Exercises

Name _____ Date _____

6.2   Trigonometry of Right Triangles

I. Trigonometric Ratios

Consider a right triangle with θ as one of its acute angles. Complete the following trigonometric ratios.

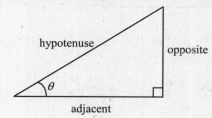

$$\sin \theta = \text{—————}$$ $$\cos \theta = \text{—————}$$ $$\tan \theta = \text{———}$$

$$\csc \theta = \text{————}$$ $$\sec \theta = \text{————}$$ $$\cot \theta = \text{————}$$

II. Special Triangles

Complete the following table of values of the trigonometric ratios for special angles.

in degrees	in radians	$\sin \theta$	$\cos \theta$	$\tan \theta$	$\csc \theta$	$\sec \theta$	$\cot \theta$
30°	$\dfrac{\pi}{6}$						
45°	$\dfrac{\pi}{4}$						
60°	$\dfrac{\pi}{3}$						

III. Applications of Trigonometry of Right Triangles

A triangle has six parts: _____. To solve a triangle

means to _____

_____.

Example 1:   Find the value of a in the given triangle.

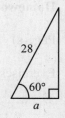

If an observer is looking at an object, then the line from the eye of the observer to the object is called _____ _____. If the object being observed is above the horizontal, then the angle between the line of sight and the horizontal is called _____. If the object is below the horizontal, then the angle between the line of sight and the horizontal is called _____ _____. If the line of sight follows a physical object, such as an inclined plane or hillside, we use the term _____.

Homework Assignment

Page(s)

Exercises

Name _____ Date _____

6.3 Trigonometric Functions of Angles

I. Trigonometric Functions of Angles

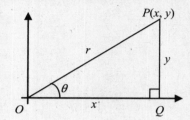

Let θ be an angle in standard position and let $P(x, y)$ be a point on the terminal side. If $r = \sqrt{x^2 + y^2}$ is the distance from the origin to the point $P(x, y)$, then the definitions of the trigonometric functions are

$\sin \theta = $ —— $\cos \theta = $ —— $\tan \theta = $ ——

$\csc \theta = $ —— $\sec \theta = $ —— $\cot \theta = $ ——

The angles for which the trigonometric functions may be undefined are the angles for which either the x- or y-coordinate of a point on the terminal side of the angle is 0. These angles are called _____,

that is, angles that are _____ with the coordinate axes.

It is a crucial fact that the values of the trigonometric functions _____ on the choice of the point $P(x, y)$.

II. Evaluating Trigonometric Functions at Any Angle

Complete the following table to indicate which trigonometric functions are positive and which are negative in each quadrant.

Quadrant	Positive Functions	Negative Functions
I		
II		
III		
IV		

Let θ be an angle in standard position. The **reference angle** $\bar{\theta}$ associated with θ is _____

_____ .

List the steps for finding the values of the trigonometric functions for any angle θ.

III. Trigonometric Identities

Complete each of the following fundamental trigonometric identities.

$\dfrac{1}{\sin\theta} = $ _____ $\dfrac{1}{\cos\theta} = $ _____ $\dfrac{1}{\tan\theta} = $ _____

$\dfrac{\sin\theta}{\cos\theta} = $ _____ $\dfrac{\cos\theta}{\sin\theta} = $ _____

$\sin^2\theta + \cos^2\theta = $ _____ $\tan^2\theta + 1 = $ _____

$1 + \cot^2\theta = $ _____

IV. Areas of Triangles

The area A of a triangle with sides of lengths a and b and with included angle θ is

$A = $ _____

Homework Assignment

Page(s)

Exercises

Name _____ Date _____

6.4   Inverse Trigonometric Functions and Right Triangles

I. The Inverse Sine, Inverse Cosine, and Inverse Tangent Functions

The sine function, on the restricted domain _____; the cosine function,
on the restricted domain _____; and the tangent function, on the restricted
domain _____; are all one-to-one and so have inverses.

The **inverse sine function** is the function $\sin^{-1}$ with domain _____ and range
_____ defined by $\sin^{-1} x = y \iff \sin y = x$.

The **inverse cosine function** is the function $\cos^{-1}$ with domain _____ and
range _____ defined by $\cos^{-1} x = y \iff \cos y = x$.

The **inverse tangent function** is the function $\tan^{-1}$ with domain _____ and range
_____ defined by $\tan^{-1} x = y \iff \tan y = x$.

The inverse sine function is also called _____, denoted by _____.

The inverse cosine function is also called _____, denoted by _____.

The inverse tangent function is also called _____, denoted by _____.

II. Solving for Angles in Right Triangles

The _____ can be used to solve for angles in a right
triangle if the lengths of at least two sides are known.

Which inverse trigonometric function should be used to find the measure of angle θ in the right triangle below?

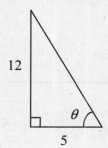

12

θ

5

III. Evaluating Expressions Involving Inverse Trigonometric Functions

Describe a possible solution strategy for evaluating an expression such as $\sin\left(\tan^{-1}\dfrac{12}{5}\right)$.

Homework Assignment

Page(s)

Exercises

Name _____   Date _____

6.5   The Law of Sines

The trigonometric functions can be used to solve oblique triangles. List the four oblique triangle cases which may be solved.

Which of these cases are solved using the Law of Sines?

I. The Law of Sines

The Law of Sines says that _____

_____ .

For a triangle *ABC*, state the Law of Sines.

II. The Ambiguous Case

The triangle situation given in Case 2, in which two sides and the angle opposite one of those sides are known, is sometimes called the **ambiguous case** because _____

_____ .

In general, if _____, we must check the angle and its supplement as possibilities, because any angle smaller than 180° can be in the triangle. To decide whether either possibility works, check to see whether the resulting sum of the angles exceeds _____. It can happen that both possibilities are compatible with the given information. In that case, _____

_____ .

Additional notes

Name _____   Date _____

6.6   The Law of Cosines

Which oblique triangle cases are solved using the Law of Cosines?

I. The Law of Cosines

The **Law of Cosines** says that in any triangle ABC, these three relationships exist among the angles A, B, and C and their opposite sides a, b, c:

In words, the Law of Cosines says that _____

_____

_____.

II. Navigation: Heading and Bearing

In navigation a direction is often given as a **bearing,** that is, as _____

_____.

III. The Area of a Triangle

Heron's Formula for the area of a triangle is an application of _____.

Heron's Formula states that the area of a triangle ABC is given by

where $s =$ _____   and is the **semiperimeter** of the triangle; that is, s is _____

_____.

Additional notes

Homework Assignment

Page(s)

Exercises

Chapter 7   Analytic Trigonometry

7.1   Trigonometric Identities

List the five trigonometric reciprocal identities.

List the three Pythagorean identities.

List three even-odd identities for trigonometry.

List the six trigonometric cofunction identities.

I. Simplifying Trigonometric Expressions

Trigonometric identities enable us to write the same expression _____. With these identities it is often possible to rewrite a complicated-looking expression as _____ _____. To simplify trigonometric expression, we use _____ _____.

II. Proving Trigonometric Identities

How can you tell if a given equation is **not** an identity?

List the guidelines for proving trigonometric identities.

Another method for proving that an equation is an identity is to transform each side of the equation separately, by way of identities, to _____. If this is possible, then _____ _____.

Homework Assignment

Page(s)

Exercises

Name _____ Date _____

7.2 Addition and Subtraction Formulas

I. Addition and Subtraction Formulas

List the addition and subtraction formulas for sine.

List the addition and subtraction formulas for cosine.

List the addition and subtraction formulas for tangent.

II. Evaluating Expressions Involving Inverse Trigonometric Functions

When evaluating expressions involving inverse trigonometric functions, remember that an expression like $\cos^{-1} x$ represents a(n) _____.

III. Expressions of the Form $A \sin x + B \cos x$

We can write expressions of the form $A \sin x + B \cos x$ in term of a single trigonometric function using _____.

If A and B are real numbers, then $A \sin x + B \cos x =$ _____, where $k = \sqrt{A^2 + B^2}$, and φ satisfies _____.

Additional notes

Homework Assignment

Page(s)

Exercises

Name _____ Date _____

7.3   Double-Angle, Half-Angle, and Product-Sum Formulas

The **Double-Angle Formulas** allow us to _____

_____.

The **Half-Angle Formulas** relate _____

_____.

The **Product-Sum Formulas** relate _____

_____.

I. Double-Angle Formulas

List the Double-Angle Formula for sine.

List the Double-Angle Formula for cosine in all its forms.

List the Double-Angle Formula for tangent.

II. Half-Angle Formulas

The Formulas for Lowering Powers allow us to write any trigonometric expression involving even powers of sine and cosine in terms of _____.

List the three Formulas for Lowering Powers.

List the three Half-Angle Formulas. Be sure to list both forms of the Half-Angle Formula for tangent.

The choice of the + or − sign depends on _____.

III. Product-Sum Formulas

List the four Product-to-Sum Formulas.

List the four Sum-to-Product Formulas.

Homework Assignment

Page(s)

Exercises

Name _____ Date _____

7.4   Basic Trigonometric Equations

A **trigonometric equation** is _____.

I. Basic Trigonometric Equations

Solving any trigonometric equation always reduces to solving a basic trigonometric equation, that is, an equation of the form _____.

To solve a basic trigonometric equation, first find the solutions in _____, and then find all solutions of the equation by adding _____
_____.

Describe how to solve a basic trigonometric equation such as $\cos\theta = 1$.

II. Solving Trigonometric Equations by Factoring

Factoring is one of the most useful techniques for solving equations, including trigonometric equations. The idea is _____
_____.

State the Zero-Product Property.

Give an example of a trigonometric equation of quadratic type.

Additional notes

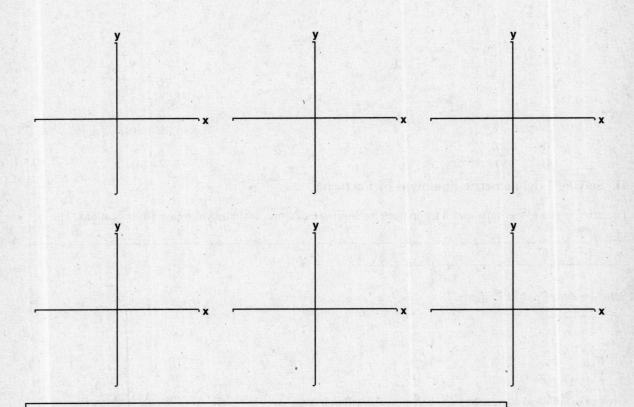

Homework Assignment

Page(s)

Exercises

Name _____ Date _____

7.5   More Trigonometric Equations

I. Solving Trigonometric Equations by Using Identities

Describe the first step in solving a trigonometric equation such as $\cos 2\theta - 5\cos\theta - 2 = 0$.

Describe a strategy for solving an equation such as $\cos\theta = 3\sin\theta + 10$.

Describe two methods of finding the values of x for which the graphs of two trigonometric functions intersect.

II. Equations with Trigonometric Functions of Multiple Angles

When solving trigonometric equations that involve functions of multiples of angles, first solve for _____

_____.

Describe how to solve an equation such as $\cos 5\theta - 1 = 0$

Additional notes

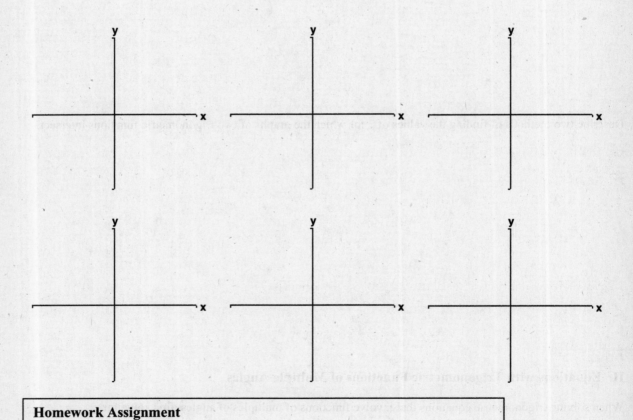

Homework Assignment

Page(s)

Exercises

Chapter 8   Polar Coordinates and Parametric Equations

8.1   Polar Coordinates

I. Definition of Polar Coordinates

The **polar coordinate system** uses distances and directions to _____

_____. To set up this system, we choose a fixed point O in the

plane called _____ and draw from O a ray (half-line) called the

_____. Then each point P can be assigned polar coordinates $P(r,\theta)$

where r is _____ and θ is _____

_____.

We use the convention that θ is positive if _____

_____ or negative if _____.

If r is negative, then $P(r,\theta)$ is defined to be _____

_____.

Because the angles $\theta + 2n\pi$ (where n is any integer) all have the same terminal side as the angle θ, each point

in the plane has _____.

II. Relationship Between Polar and Rectangular Coordinates

To change from polar to rectangular coordinates, use the formulas

To change from rectangular to polar coordinates, use the formulas

III. Polar Equations

To convert an equation from rectangular to polar coordinates, simply _____

_____.

Describe one or more strategies for converting a polar equation to rectangular form.

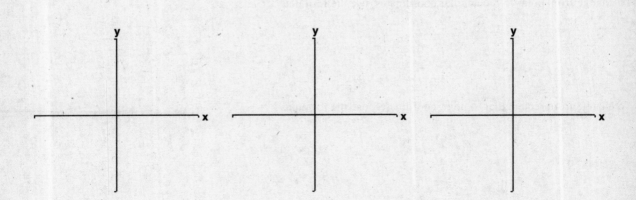

Name _____ Date _____

8.2   Graphs of Polar Equations

The **graph of a polar equation** $r = f(\theta)$ consists of _____

_____.

I. Addition and Subtraction Formulas

To plot points in polar coordinates, it is convenient to use a grid consisting of _____

_____.

The graph of the equation $r = a$ is _____.

To sketch a polar curve whose graph isn't obvious, _____

_____.

The graphs of equations of the form $r = 2a \sin \theta$ and $r = 2a \cos \theta$ are _____

_____.

A **cardioid** has the shape of _____. The graph of any equation of the form

_____ or _____ is a cardioid.

The graph of an equation of the form $r = a \cos n\theta$ or $r = a \sin n\theta$ is an n-leaved rose if _____ or

a $2n$-leaved rose if _____.

II. Symmetry

In graphing a polar equation, it's often helpful to take advantage of _____.

List three tests for symmetry.

In polar coordinates, the zeros of the function $r = f(\theta)$ are the angles θ at which _____

_____ .

The graph of an equation of the form $r = a \pm b\cos\theta$ or $r = a \pm b\sin\theta$ is a _____ .

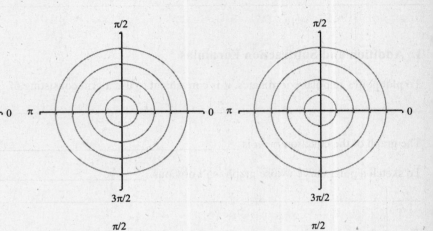

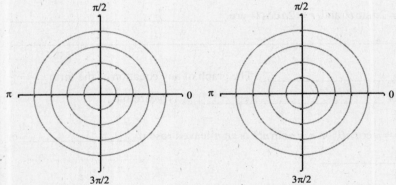

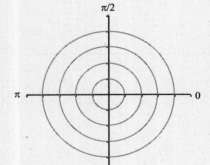

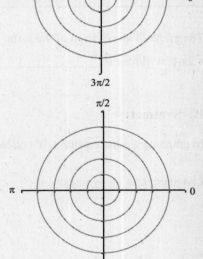

Homework Assignment

Page(s)

Exercises

Name _____ Date _____

8.3 Polar Form of Complex Numbers; De Moivre's Theorem

I. Graphing Complex Numbers

How many axes are needed to graph a complex number? Explain.

The complex plane is determined by the _____ and the _____.

To graph the complex number $a + bi$ in the complex plane, plot the ordered pair of numbers _____ in this plane.

The **modulus**, or **absolute value**, of the complex number $z = a + bi$ is _____.

II. Polar Form of Complex Numbers

A complex number $z = a + bi$ has the **polar form,** or **trigonometric form,** _____,

where $r =$ _____ and $\tan \theta =$ _____ . The number r

is the _____ of z, and θ is an _____ of z.

The argument of z is not unique, but any two arguments of z differ by a _____.
When determining the argument, we must consider _____.

If the two complex numbers z_1 and z_2 have the polar forms $z_1 = r_1(\cos\theta_1 + i\sin\theta_1)$ and
$z_2 = r_2(\cos\theta_2 + i\sin\theta_2)$, then the numbers are multiplied and divided as follows.

$z_1 z_2 =$ _____

$\dfrac{z_1}{z_2} =$ _____

This theorem says that to multiply two complex numbers, _____

_____ . It also says that to divide complex numbers, _____

_____ .

III. De Moivre's Theorem

De Moivre's Theorem gives a useful formula for _____

_____ .

De Moivre's Theorem states that if $z = r(\cos\theta + i\sin\theta)$, then for any integer n,

$z^n = $ _____.

Give an interpretation of De Moivre's Theorem.

IV. *n*th Roots of Complex Numbers

An ***n*th root** of a complex number z is _____.

If $z = r(\cos\theta + i\sin\theta)$ and n is a positive integer, then z has the n distinct nth roots

_____, for $k = 0, 1, 2, \ldots, n-1$.

When finding the nth roots of $z = r(\cos\theta + i\sin\theta)$, notice that the modulus of each nth root is _____.
Also, the argument of the first root is _____. Furthermore, we repeatedly add _____ to get the argument of each successive root.

When graphed, the nth roots of z are spaced equally on _____.

Homework Assignment

Page(s)

Exercises

Name _____ Date _____

8.4 Plane Curves and Parametric Equations

I. Plane Curves and Parametric Equations

We can think of a curve as the path of a point moving in the plane; the x-coordinates and y-coordinates of the point are then _____.

If f and g are functions defined on an interval I, then the set of points $(f(t), g(t))$ is a _____.
The equations $x = f(t)$ and $y = g(t)$ where $t \in I$, are _____ for the
curve, with **parameter** t.

A parametrization contains more information than just the shape of the curve; it also indicates _____
_____.

II. Eliminating the Parameter

Often a curve given by parametric equations can also be represented by a single rectangular equation in x and y.
The process of finding this equation is called _____. One way
to do this is _____.

To identify the shape of a parametric curve, _____
_____.

III. Finding Parametric Equations for a Curve

Describe how to find a set of parametric equations for a curve.

A **cycloid** is _____
_____.

Name two interesting physical properties of the cycloid.

IV. Using Graphing Devices to Graph Parametric Curves

A **closed curve** is _____.

A **Lissajous figure** is the graph of a pair of parametric equations of the form _____ and
_____ where A, B, ω_1, and ω_2 are real constants.

The graph of the polar equation $r = f(\theta)$ is the same as the graph of the parametric equations
_____ and _____.

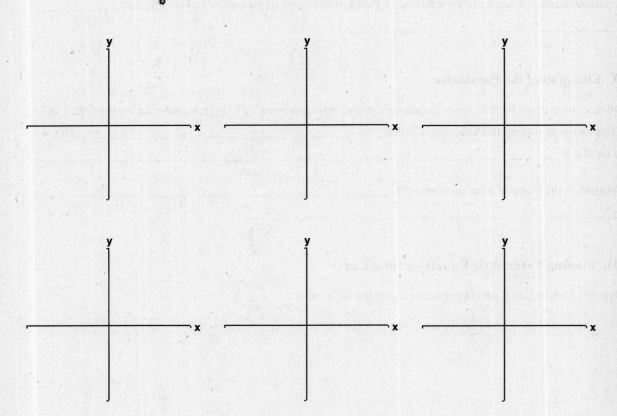

Homework Assignment

Page(s)

Exercises

Chapter 9   Vectors in Two and Three Dimensions

9.1   Vectors in Two Dimensions

A **scalar** is _____.

Quantities such as displacement, velocity, acceleration, and force that involve magnitude as well as direction are called _____. One way to represent such quantities mathematically is through the use of _____.

I. Geometric Description of Vectors

A **vector** in the plane is _____. We sketch a vector as _____. If a vector between points A and B is denoted as $\overrightarrow{AB}$, then point A is the _____, and point B is the _____ _____. The length of the line segment AB is called the _____ of the vector and is denoted by $\left|\overrightarrow{AB}\right|$. We use _____ letters to denote vectors.

Two vectors are considered **equal** if they have _____.

If the displacement $\mathbf{u} = \overrightarrow{AB}$ is followed by the displacement $\mathbf{v} = \overrightarrow{BC}$, then the resulting displacement is _____. In other words, the single displacement represented by the vector $\overrightarrow{AC}$ has the same effect as _____. We call the vector $\overrightarrow{AC}$ the _____ of the vectors $\overrightarrow{AB}$ and $\overrightarrow{BC}$, and we write _____. The **zero vector**, denoted by $\mathbf{0}$, represents _____. Thus to find the sum of any two vectors $\mathbf{u}$ and $\mathbf{v}$, _____ _____.

If we draw $\mathbf{u}$ and $\mathbf{v}$ starting at the same point, then $\mathbf{u} + \mathbf{v}$ is the vector that is _____ _____.

Describe the process of **multiplication of a vector by a scalar** and the effect it has on the vector.

The **difference** of two vectors **u** and **v** is defined by _____.

II. Vectors in the Coordinate Plane

In the coordinate plane, we represent **v** as an ordered pair of real numbers **v** = _____, where a
is the _____ and b is the _____
_____.

If a vector **v** is represented in the plane with initial point $P(x_1, y_1)$ and terminal point $Q(x_2, y_2)$, then

v = _____.

The **magnitude** or **length** of a vector **v** = $\langle a, b \rangle$ is _____.

If **u** = $\langle a_1, b_1 \rangle$ and **v** = $\langle a_2, b_2 \rangle$, then

u + **v** = _____

u − **v** = _____

c**u** = _____

The **zero vector** is the vector _____.

List four properties of vector addition.

List the property for the length of a vector.

List six properties of multiplication by a scalar.

A vector of length 1 is called a _____. Two useful unit vectors are **i** and **j**, defined by **i** = _____ and **j** = _____.

The vector $\mathbf{v} = \langle a, b \rangle$ can be expressed in terms of **i** and **j** by **v** = _____.

Let **v** be a vector in the plane with its initial point at the origin. The **direction** of **v** is θ, _____ _____.

Horizontal and Vertical Components of a Vector

Let **v** be a vector with magnitude | **v** | and direction θ. Then $\mathbf{v} = \langle a, b \rangle = a\mathbf{i} + b\mathbf{j}$, where a = _____, and b = _____. Thus we can express **v** as _____.

III. Using Vectors to Model Velocity and Force

The **velocity** of a moving object is modeled by _____ _____.

Force is also represented by a vector. We can think of force as _____ _____. Force is measured in _____. If several forces are acting on an object, the **resultant force** experienced by the object is _____ _____.

Additional notes

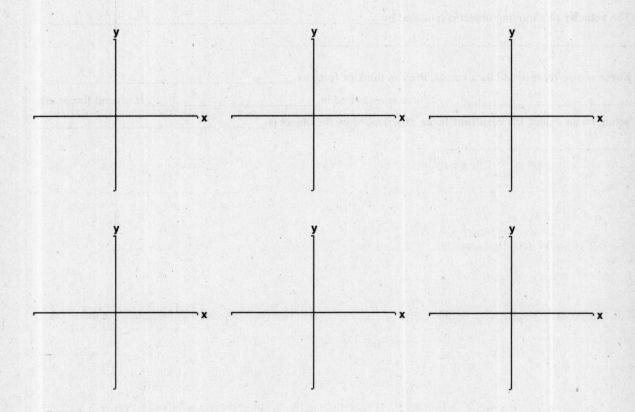

```
Homework Assignment

Page(s)

Exercises
```

Name _____ Date _____

9.2   The Dot Product

I. The Dot Product of Vectors

If $\mathbf{u} = \langle a_1, b_1 \rangle$ and $\mathbf{v} = \langle a_2, b_2 \rangle$ are vectors, then their dot product, denoted by $\mathbf{u} \cdot \mathbf{v}$, is defined by $\mathbf{u} \cdot \mathbf{v} =$

_____. Thus to find the dot product of $\mathbf{u}$ and $\mathbf{v}$, we _____

_____.

The dot product of $\mathbf{u}$ and $\mathbf{v}$ is *not* a vector; it is a _____.

List four properties of the dot product.

Let $\mathbf{u}$ and $\mathbf{v}$ be vectors, and sketch them with initial points at the origin. We define the **angle θ between u and v**

to be _____,

thus $0 \le \theta \le \pi$.

The **Dot Product Theorem** states that if θ is the angle between two nonzero vectors $\mathbf{u}$ and $\mathbf{v}$, then

_____.

The Dot Product Theorem is useful because it allows us to find the angle between two vectors if we know the

components of the vectors. If θ is the angle between two nonzero vectors $\mathbf{u}$ and $\mathbf{v}$, then $\cos \theta =$ _____.

Two nonzero vectors $\mathbf{u}$ and $\mathbf{v}$ are called **perpendicular,** or **orthogonal,** if the angle between them is _____.

We can determine whether two vectors are perpendicular by finding their dot product. Two nonzero vectors $\mathbf{u}$

and $\mathbf{v}$ are perpendicular if and only if _____.

II. The Component of u Along v

The **component of u along v** (or the **component of u in the direction of v**) is defined to be _____,
where θ is the angle between **u** and **v**. Intuitively, the component of **u** along **v** is the magnitude of _____
_____.

The component of **u** along **v** is calculated as _____ .

_____

III. The Projection of u onto v

The projection of **u** onto **v**, denoted by proj$_v$ **u**, is _____
_____. The projection of **u** onto **v** is given by

proj$_v$ **u** = _____ . If the vector **u** is **resolved** into u_1 and u_2, where u_1 is parallel to **v** and

_____

u_2 is orthogonal to **v**, then $u_1 =$ _____ and $u_2 =$ _____ .

IV. Work

One use of the dot product occurs in calculating _____.

The work W done by a force **F** in moving along a vector **D** is _____.

Homework Assignment

Page(s)

Exercises

Name _____ Date _____

9.3 Three-Dimensional Coordinate Geometry

I. The Three-Dimensional Rectangular Coordinate System

To represent points in space, we first choose a fixed point O (the origin) and three directed lines through O that are perpendicular to each other, called the _____ and labeled the

_____. We usually think of the _____ as being horizontal and the _____ as being vertical.

The three coordinate axes determine the three _____. The _____ is the plane that contains the x- and y-axes. The _____ is the plane that contains the y- and z-axes. The _____ is the plane that contains the x- and z-axes. These three coordinate planes divide space into eight parts called _____.

Any point P in space can be located by a unique _____ of real numbers (a, b, c). The first number a is _____, the second number b is _____ _____, and the third number c is _____. The set of all ordered triples $\{(x, y, z) \mid x, y, z \in \mathbb{R}\}$ forms the _____

_____.

In three-dimensional geometry, an equation in x, y, and z represents a three-dimensional _____.

II. Distance Formula in Three Dimensions

The distance between the points $P(x_1, y_1, z_1)$ and $Q(x_1, y_1, z_1)$ is

_____

Example 1 Find the distance between the points $(1, 3, -5)$ and $(2, 1, 0)$.

III. The Equation of a Sphere

An equation of a sphere with center $C(h, k, l)$ and radius r is _____.

Find the radius and center of the sphere with equation $(x+2)^2 + (y-9)^2 + (z+14)^2 = 25$.

The intersection of a sphere with a plane is called the _____ of the sphere in a plane.

Homework Assignment

Page(s)

Exercises

Name _____ Date _____

9.4　Vectors in Three Dimensions

I. Vectors in Space

If a vector **a** is represented in space with initial point $P(x_1, y_1, z_1)$ and terminal point $Q(x_2, y_2, z_2)$, then the

component form of a vector in space is given as **a** = _____.

The magnitude of the three-dimensional vector $\mathbf{a} = \langle a_1, a_2, a_3 \rangle$ is _____.

II. Combining Vectors in Space

If $\mathbf{a} = \langle a_1, a_2, a_3 \rangle$, $\mathbf{b} = \langle b_1, b_2, b_3 \rangle$, and c is a scalar, then complete each of the following algebraic operations on vectors in three dimensions.

a + **b** = _____

a − **b** = _____

c**a** = _____

The three-dimensional vectors $\mathbf{i} = \langle 1, 0, 0 \rangle$, $\mathbf{j} = \langle 0, 1, 0 \rangle$, and $\mathbf{k} = \langle 0, 0, 1 \rangle$ are examples of _____.

The vector $\mathbf{a} = \langle a_1, a_2, a_3 \rangle$ can be expressed in terms of **i**, **j**, and **k** by $\mathbf{a} = \langle a_1, a_2, a_3 \rangle$ = _____.

III. The Dot Product for Vectors in Space

If $\mathbf{a} = \langle a_1, a_2, a_3 \rangle$ and $\mathbf{b} = \langle b_1, b_2, b_3 \rangle$ are vectors in three dimensions, then their **dot product** is defined by

_____.

Let **u** and **v** be vectors in space and θ be the angle between them. Then $\cos \theta =$ _____ .

In particular, **u** and **v** are **perpendicular** (or **orthogonal**) if and only if _____.

IV. Direction Angles of a Vector

The **direction angles** of a nonzero vector $\mathbf{a} = a_1\mathbf{i} + a_2\mathbf{j} + a_3\mathbf{k}$ are the angles α, β, and γ in the interval $[0, \pi]$ that

_____ .

The cosines of these angles, $\cos\alpha$, $\cos\beta$, and $\cos\gamma$, are called the _____

_____ . If $\mathbf{a} = a_1\mathbf{i} + a_2\mathbf{j} + a_3\mathbf{k}$ is a nonzero vector in space, the direction angles α, β, and

γ satisfy

_____

If $|\mathbf{a}| = 1$, then the direction cosines of $\mathbf{a}$ are simply _____ .

The direction angles α, β, and γ of a nonzero vector $\mathbf{a}$ in space satisfy the following equation

_____

This property indicates that if we know two of the direction cosines of a vector, we can find _____

_____ .

Homework Assignment

Page(s)

Exercises

Name _____ Date _____

9.5 The Cross Product

I. The Cross Product

If $a = \langle a_1, a_2, a_3 \rangle$ and $b = \langle b_1, b_2, b_3 \rangle$ are three-dimensional vectors, then the **cross product** of **a** and **b** is the

vector _____.

The cross product $a \times b$ of two vectors **a** and **b**, unlike the dot product, is a _____.

Note that $a \times b$ is defined only when **a** and **b** are vectors in _____.

To help us remember the definition of the cross product, we use the notation of _____.
A **determinant of order three** is defined in terms of second-order determinants as

_____

Each term on the right side of the third-order determinant equation involves a number a_i in the first row of the

determinant, and a_i is multiplied by the second-order determinant obtained from the left side by _____
_____

We can write the definition of the cross-product using determinants as

_____

II. Properties of the Cross Product

The Cross Product Theorem states that the vector $a \times b$ is _____ to both
a and **b**.

If **a** and **b** are represented by directed line segments with the same initial point, then the Cross Product Theorem
says that the cross product $a \times b$ points _____
_____. It turns out that the direction of $a \times b$ is given by the *right-
hand rule:* _____
_____

If θ is the angle between **a** and **b** (so $0 \leq \theta \leq \pi$), then the length of the cross product of **a** and **b** is given by

_____

In particular, two nonzero vectors **a** and **b** are parallel if and only if _____.

III. Area of a Parallelogram

The length of the cross product $\mathbf{a} \times \mathbf{b}$ is the area of _____

_____.

IV. Volume of a Parallelepiped

The product $\mathbf{a} \cdot (\mathbf{b} \times \mathbf{c})$ is called the _____ of the vectors **a**, **b**, and **c**.

The scalar triple product can be written as the following determinant

_____

The volume of the parallelepiped, a three-dimensional figure having parallel faces, determined by the vectors **a**, **b**, and **c** is the magnitude of their scalar triple product _____. In particular, if the volume of the parallelepiped is 0, then the vectors **a**, **b**, and **c** are _____.

Homework Assignment

Page(s)

Exercises

Name _____ Date _____

9.6   Equations of Lines and Planes

I. Equations of Lines

A line L in three-dimensional space is determined when we know a point $P_0(x_0, y_0, z_0)$ on L and _____

_____. In three dimensions the direction of a line is described by _____

_____. The line L is given by the position vector $\mathbf{r}$, where $\mathbf{r} =$ _____ for

$t \in \mathbb{R}$, and $\mathbf{r}_0$ is the position vector of P_0. This the **vector** _____.

A line passing through the point $P(x_0, y_0, z_0)$ and parallel to the vector $\mathbf{v} = \langle a, b, c \rangle$ is described by the parametric equations

_____

where t is any real number.

II. Equations of Planes

Although a line in space is determined by a point and a direction, the "direction" of a plane cannot _____

_____. In fact, different vectors in a plane can have

different directions. But a vector _____ to a plane *does* completely specify the

direction of the plane. Thus a plane in space is determined by _____

_____. This orthogonal vector $\mathbf{n}$ is

called a _____.

The plane containing the point $P(x_0, y_0, z_0)$ and having the normal vector $\mathbf{n} = \langle a, b, c \rangle$ is described by the

equation _____.

Additional notes

Homework Assignment

Page(s)

Exercises

Chapter 10   Systems of Equations and Inequalities

10.1   Systems of Linear Equations in Two Variables

I. Systems of Linear Equations and Their Solutions

A **system of equations** is _____.

A **system of linear equations** is _____.

A **solution** of a system is _____

_____. To **solve** a system means to _____

_____.

II. Substitution Method

In the **substitution method,** we start with _____

_____.

Describe the substitution method procedure.

III. Elimination Method

To solve a system using the **elimination method,** we try to _____

_____.

Describe the elimination method procedure.

IV. Graphical Method

In the **graphical method,** we use _____.

Describe the graphical method procedure.

V. The Number of Solutions of a Linear System in Two Variables

The graph of a linear system in two variables is _____, so to solve the system
graphically, we must find _____.

For a system of linear equations in two variables, exactly one of the following is true concerning the number of
solutions the system has.

1.

2.

3.

A system that has no solution is said to be _____. A system with infinitely
many solutions is called _____.

If solving a system of equations eliminates both variables and results in a statement which is false, such as
$0 = 7$, then the system has _____.

VI. Modeling with Linear Systems

State the guidelines for modeling with systems of equations.

Homework Assignment
Page(s)
Exercises

Name _____ Date _____

10.2   Systems of Linear Equations in Several Variables

A **linear equation in *n* variables** is _____
_____.

I. Solving a Linear System

A linear system in the three variables x, y, and z is in **triangular form** if _____
_____
_____.

It is easy to solve a system that is in triangular form by using _____.

To change a system of linear equations to an equivalent system, that is, a system with _____
_____, we use the _____.

List the operations that yield an equivalent system.

To solve a linear system, we use these operations to change the system to an equivalent _____
_____, and then use back-substitution to complete the solution. This process is called _____
_____.

II. The Number of Solutions of a Linear System

The graph of a linear equation in three variables is _____.

A system of three equations in three variables represents _____. The

solutions of the system are _____.

For a system of linear equations, exactly one of the following is true concerning the number of solutions the
system has.

1.

2.

3.

A system with no solutions is said to be _____, and a system with infinitely many solutions is said to be _____. A linear system has no solution if we end up with _____ after applying Gaussian elimination to the system.

III. Modeling Using Linear Systems

Linear systems are used to model situations that involve _____.

Homework Assignment

Page(s)

Exercises

Name _____ Date _____

10.3 Matrices and Systems of Linear Equations

A matrix is simply _____. Matrices are used to
organize information into categories that correspond to _____
_____.

I. Matrices

An $m \times n$ **matrix** is a rectangular array of numbers with _____ **rows** and _____ **columns**.
We say that the matrix has _____ $m \times n$. The numbers a_{ij} are the _____ of
the matrix. The subscript on the entry a_{ij} indicates that it is in _____.

Example 1   Give the dimension of the following matrix.
$$\begin{bmatrix} -1 & 3 & 0 & 1 & 5 \\ 2 & -3 & -5 & 4 & -2 \\ 0 & -4 & 1 & -1 & 4 \end{bmatrix}$$

II. The Augmented Matrix of a Linear System

We can write a system of linear equations as a matrix, called the **augmented matrix** of the system, by
_____.

Example 2   Write the augmented matrix for the linear system $\begin{cases} 2x + 15y - 3z = 12 \\ 4x - 3y + 15z = -3 \\ -x + 10y + 5z = 10 \end{cases}$.

III. Elementary Row Operations

List the elementary row operations.

Performing any of these operations on the augmented matrix of a system _____ its solution.

Give a description of each of the following notations that are used to represent elementary row operations.

$R_i + kR_j \rightarrow R_i$ _____

kR_i _____

$R_i \leftrightarrow R_j$ _____

IV. Gaussian Elimination

To solve a system of linear equations using its augmented matrix, _____
_____.

List the conditions for which a matrix is considered to be in **row-echelon form**.

A matrix is in **reduced row-echelon form** if it is in row-echelon form and also satisfies the condition that
_____.

Describe a systematic way to put a matrix in row-echelon form using elementary row operations.

Once an augmented matrix is in row-echelon form, the corresponding linear system may be solved _____
_____. This technique is called _____.

List the steps for solving a system using Gaussian elimination.

V. Gauss-Jordan Elimination

If we put the augmented matrix of a linear system in *reduced* row-echelon form, then we don't need to _____
_____.

List the steps for putting a matrix in reduced row-echelon form.

Using the reduced row-echelon form to solve a system is called _____.

VI. Inconsistent and Dependent Systems

A **leading variable** in a linear system is one that _____
_____.

Suppose the augmented matrix of a system of linear equations has been transformed by Gaussian elimination into row-echelon form. Then exactly one of the following is true concerning the number of solutions the system has.

A system with no solution is called _____.

If a system in row-echelon form has n nonzero equations in m variables ($m > n$), then the complete solution will have _____ nonleading variables.

Additional notes

Name _____ Date _____

10.4 The Algebra of Matrices

I. Equality of Matrices

Two matrices are equal if _____.

The formal definition of matrix equality states that matrices $A = \left[a_{ij} \right]$ and $B = \left[b_{ij} \right]$ are **equal** if and only if

they have _____, and corresponding entries are equal, that is

_____, that is, for $i = 1, 2, \ldots, m$ and $j = 1, 2, \ldots, n$.

II. Addition, Subtraction, and Scalar Multiplication of Matrices

Let $A = \left[a_{ij} \right]$ and $B = \left[b_{ij} \right]$ be matrices of the same dimension $m \times n$, and let c be any real number.

1. The **sum** $A + B$ is the $m \times n$ matrix obtained by _____.

   $A + B =$ _____

2. The **difference** $A - B$ is the $m \times n$ matrix obtained by _____

   _____.

   $A - B =$ _____

3. The **scalar product** cA is the $m \times n$ matrix obtained by _____.

   $cA =$ _____

Let A, B, and C be $m \times n$ matrices and let c and d be scalars. State each of the following properties of matrix arithmetic.

Commutative Property of Matrix Addition: _____

Associative Property of Matrix Addition: _____

Associative Property of Scalar Multiplication: _____

Distributive Properties of Scalar Multiplication: _____

_____

III. Multiplication of Matrices

The product AB of two matrices A and B is defined only when _____

_____. This means that if we write their dimensions side by

side, _____. If this is true, then the product

AB is a matrix of dimension _____.

If $\begin{bmatrix} a_1 & a_2 & \cdots & a_n \end{bmatrix}$ is a row of A, and if $\begin{bmatrix} b_1 \\ b_2 \\ \vdots \\ b_n \end{bmatrix}$ is a column of B, then their **inner product** is the number

_____.

The definition of matrix multiplication states that if $A = \begin{bmatrix} a_{ij} \end{bmatrix}$ is an $m \times n$ matrix and $B = \begin{bmatrix} b_{ij} \end{bmatrix}$ is an $n \times k$

matrix, then their product is the _____, where c_{ij} is the inner product

of the ith row of A and the jth column of B. We write the product as _____.

The definition of matrix product says that each entry in the matrix AB is obtained from a _____ of A

and a _____ of B as follows: _____

_____

_____.

Example 1   Suppose A is a 2×6 matrix and B is a 4×2 matrix. Which of the following products is
possible: $A \times B$, $B \times A$, both of these, or neither of these?

Example 2   Consider the matrix product that is possible in Example 1. What is the dimension of the
possible product(s)?

IV. Properties of Matrix Multiplication

Let A, B, and C be matrices for which the following products are defined. List the properties of matrix multiplication that are true.

Matrix multiplication _____ commutative.

V. Applications of Matrix Multiplication

Give an example of an application of matrix multiplication.

A matrix in which the entries of each column add up to 1 is called _____.

VI. Computer Graphics

Briefly describe how matrices are used in the digital representation of images.

Additional notes

Name _____   Date _____

10.5   Inverses of Matrices and Matrix Equations

I. The Inverse of a Matrix

The **identity matrix** I_n is the $n \times n$ matrix for which _____

_____.

Identity matrices behave like _____ in the sense that $A \cdot I_n = A$ and $I_n \cdot B = B$,

whenever these products are defined.

If A and B are $n \times n$ matrices, and if $AB = BA = I_n$, then we say that B is _____,

and we write _____.

Let A be a square $n \times n$ matrix. The definition of the inverse of a matrix states that if there exists an $n \times n$

matrix A^{-1} with the property that $AA^{-1} = A^{-1}A = I_n$, then we say that A^{-1} is the _____ of A.

II. Finding the Inverse of a 2 × 2 Matrix

The following rule provides a simple way for finding the inverse of a 2×2 matrix, when it exists.

If $A = \begin{bmatrix} a & b \\ c & d \end{bmatrix}$, then $A^{-1} =$

_____

If _____, then A has no inverse.

The quantity $ad - bc$ that appears in the rule for calculating the inverse of a 2×2 matrix is called the

_____ of the matrix. If the determinant is 0, then _____

_____.

III. Finding the Inverse of an $n \times n$ Matrix

Describe the procedure for finding the inverse of a 3 × 3 or larger matrix.

Graphing calculators are also able to calculate matrix inverses. List the steps required to do so for your graphing calculator.

If we encounter a row of zeros on the left when trying to find an inverse, then _____
_____. A matrix that has no inverse is called _____.

IV. Matrix Equations

For the system of linear equations $\begin{cases} a_1x + b_1y + c_1z = d_1 \\ a_2x + b_2y + c_2z = d_2 \\ a_3x + b_3y + c_3z = d_3 \end{cases}$, write the corresponding matrix equation, in the form

$AX = B$, where the matrix A is the **coefficient matrix,** X is the variable matrix, and B is the constant matrix.

If A is a square $n \times n$ matrix that has an inverse A^{-1} and if X is a variable matrix and B a known matrix, both with n rows, then the solution of the matrix equation $AX = B$ is given by _____.

Homework Assignment

Page(s)

Exercises

Name _____ Date _____

10.6 Determinants and Cramer's Rule

A **square** matrix is one that has _____.

I. Determinant of a 2 × 2 Matrix

We denote the determinant of a square matrix A by the symbol _____ or _____.

If $A = [a]$ is a 1×1 matrix, then $\det(A) = $ _____.

The **determinant** of the 2×2 matrix $A = \begin{bmatrix} a & b \\ c & d \end{bmatrix}$ is $\det(A) = |A| = $ _____.

II. Determinant of an $n \times n$ Matrix

Let A be an $n \times n$ matrix.

The **minor** M_{ij} of the element a_{ij} is _____

_____.

The **cofactor** A_{ij} of the element a_{ij} is _____.

Note that the cofactor of a_{ij} is simply the minor of a_{ij} multiplied by either 1 or -1, depending on whether $i + j$ is _____.

Fill in the + and – sign pattern associated the minors of a 4×4 matrix.

$$\begin{bmatrix} & & & \\ & & & \\ & & & \\ & & & \end{bmatrix}$$

If A is an $n \times n$ matrix, then the **determinant** of A is obtained by _____

_____. In symbols, this is

$\det(A) = $ _____

This definition of determinant used the cofactors of elements in the first row only. This is called _____

_____. In fact, we can expand the determinant by any row or column in

the same way and obtain _____.

The Invertibility Criterion states that if A is a square matrix, then A has an inverse if and only if

_____.

III. Row and Column Transformations

If we expand a determinant about a row or column that contains many zeros, our work is reduced considerably

because _____.

The principle of row and column transformations of a determinant states that if A is a square matrix and if the

matrix B is obtained from A by adding a multiple of one row to another or a multiple of one column to another,

then _____.

This principle often simplifies the process of finding a determinant by _____

_____.

IV. Cramer's Rule

Cramer's Rule for Systems in Two Variables states that the linear system $\begin{cases} ax + by = r \\ cx + dy = s \end{cases}$ has the solution

$x =$ _____   and $y =$ _____, provided that $\begin{vmatrix} a & b \\ c & d \end{vmatrix} \neq 0$.

For the linear system $\begin{cases} ax + by = r \\ cx + dy = s \end{cases}$, complete the notation for each of the following and give a description.

$D = \begin{bmatrix} & \\ & \end{bmatrix}$

$D_x = \begin{bmatrix} & \\ & \end{bmatrix}$

$D_y = \begin{bmatrix} & \\ & \end{bmatrix}$

Write the solution of the system $\begin{cases} ax + by = r \\ cx + dy = s \end{cases}$ using D, D_x, and D_y.

Cramer's Rule states that if a system of n linear equations in the n variables $x_1, x_2, \ldots, x_n$ is equivalent to the matrix equation $DX = B$, and if $|D| \neq 0$, then its solutions are $x_1 = \dfrac{|D_{x_1}|}{|D|}$, $x_2 = \dfrac{|D_{x_2}|}{|D|}$, $\ldots$, $x_n = \dfrac{|D_{x_n}|}{|D|}$, where

D_{x_i} is _____.

V. Areas of Triangles Using Determinants

If a triangle in the coordinate plane has vertices (a_1, b_1), (a_2, b_2), and (a_3, b_3), then its area is

Area =

————————————————

where the sign is chosen to make the area positive.

Additional notes

Name _____ Date _____

10.7 Partial Fractions

Some applications in calculus require expressing a fraction as the sum of simpler fractions called _____

_____.

Let r be the rational function $r(x) = \dfrac{P(x)}{Q(x)}$ where the degree of P is less than the degree of Q. After we have

completely factored the denominator Q of r, we can express $r(x)$ as a sum of partial fractions of the form

_____ and _____. This sum is called the _____

_____ of r.

I. Distinct Linear Factors

Case 1: The Denominator is a Product of Distinct Linear Factors

Suppose that we can factor $Q(x)$ as $Q(x) = (a_1 x + b_1)(a_2 x + b_2) \cdots (a_n x + b_n)$ with no factor repeated. In this

case the partial fraction decomposition of $P(x)/Q(x)$ takes the form

$$\frac{P(x)}{Q(x)} = \underline{\hspace{6cm}}$$

In your own words, describe the process of finding a partial fraction decomposition.

II. Repeated Linear Factors

Case 2: The Denominator is a Product of Distinct Linear Factors, Some of Which Are Repeated

Suppose the complete factorization of $Q(x)$ contains the linear factor $ax + b$ repeated k times; that is, $(ax + b)^k$

is a factor of $Q(x)$. Then, corresponding to each such factor, the partial fraction decomposition for $P(x)/Q(x)$

contains _____.

III. Irreducible Quadratic Factors

Case 3: The Denominator Has Irreducible Quadratic Factors, None of Which Is Repeated

Suppose the complete factorization of $Q(x)$ contains the quadratic factor $ax^2 + bx + c$, which can't be factored further. Then, corresponding to this, the partial fraction decomposition of $P(x)/Q(x)$ will have a term of the

form _____.

IV. Repeated Irreducible Quadratic Factors

Case 4: The Denominator Has a Repeated Irreducible Quadratic Factor

Suppose the complete factorization of $Q(x)$ contains the factor $(ax^2 + bx + c)^k$, where $ax^2 + bx + c$ can't be factored further. Then the partial fraction decomposition of $P(x)/Q(x)$ will have the terms

_____

The techniques described in this section apply only to rational functions $P(x)/Q(x)$ in which _____

_____. If this isn't the case, we must first _____

_____.

Homework Assignment

Page(s)

Exercises

Name _____ Date _____

10.8 Systems of Nonlinear Equations

I. Substitution and Elimination Methods

To solve a system of nonlinear equations, we can use the _____.

Describe the process for solving a system of nonlinear equations using the substitution method.

Describe the process for solving a system of nonlinear equations using the elimination method.

II. Graphical Method

Describe the process for solving a system of nonlinear equations with the graphical method.

Additional notes

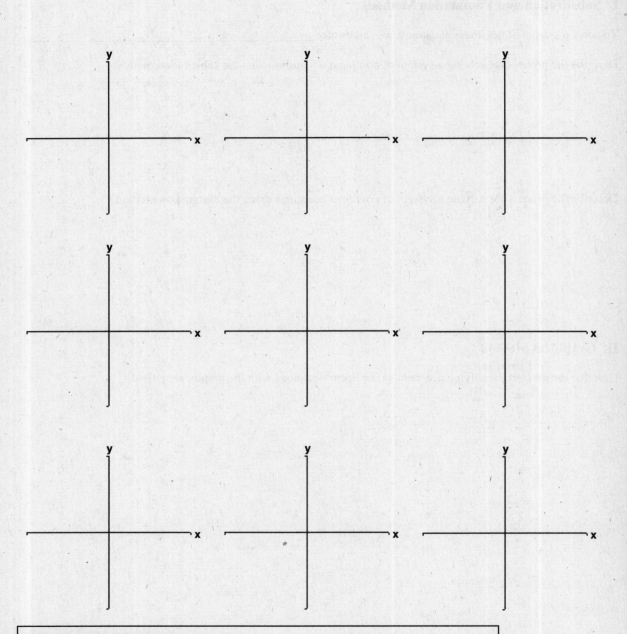

Homework Assignment

Page(s)

Exercises

Name _____ Date _____

10.9 Systems of Inequalities

I. Substitution and Elimination Methods

The graph of an inequality, in general, consists of a region in the plane whose boundary is _____
_____ .

To determine which side of the graph gives the solution set of the inequality, we need only _____
_____ .

List the steps for graphing an inequality.

II. Systems of Inequalities

The solution of a system of inequalities is the set of all points in the coordinate plane that _____
_____ .

For a system of inequalities, the vertices of the solution set are obtained by _____
_____
_____ .

III. Systems of Linear Inequalities

If an inequality can be put into one of the forms $ax + by \geq c$, $ax + by \leq c$, $ax + by > c$, or $ax + by < c$, then the

inequality is _____ .

Describe the process of solving a system of linear inequalities.

When a region in the plane can be covered by a sufficiently large circle, it is said to be _____.

A region that is not bounded is called _____. An unbounded region cannot be "fenced

in"—it extends _____.

IV. Application: Feasible Regions

Give an example of variable constraints in applied problems.

Constraints or limitations such as these can usually be expressed as _____.

When dealing with applied inequalities, we usually refer to the solution set of a system as a _____

_____, because the points in the solution set represent _____

_____.

Homework Assignment

Page(s)

Exercises

Name _____ Date _____

Chapter 11   Conic Sections

11.1   Parabolas

I. Geometric Definition of a Parabola

A **parabola** is the set of points in the plane that are _____

_____.

The **vertex** V of the parabola lies _____,

and the **axis of symmetry** is the line that runs _____

_____.

II. Equations and Graphs of Parabolas

The graph of the equation $x^2 = 4py$ is a parabola with _____ axis. Its vertex is _____,

its focus is _____, and its directrix is _____. The parabola opens upward if

_____ or downward if _____.

The graph of the equation $y^2 = 4px$ is a parabola with _____ axis. Its vertex is

_____, its focus is _____, and its directrix is _____. The parabola opens

to the right if _____ or to the left if _____.

Describe how to find the focus and directrix of a parabola from its equations.

The line segment that runs through the focus of a parabola perpendicular to the axis, with endpoints on the

parabola, is called the _____, and its length is the _____ of

the parabola. The focal diameter of a parabola is _____.

III. Applications

Parabolas have the important property that light from a source placed at the focus of a surface with parabolic

cross section will be reflected in such a way that it _____.

This property makes parabolas very useful as _____.

This reflection property of parabolas also means that light approaching a parabolic reflector in rays parallel to its axis of symmetry is _____.

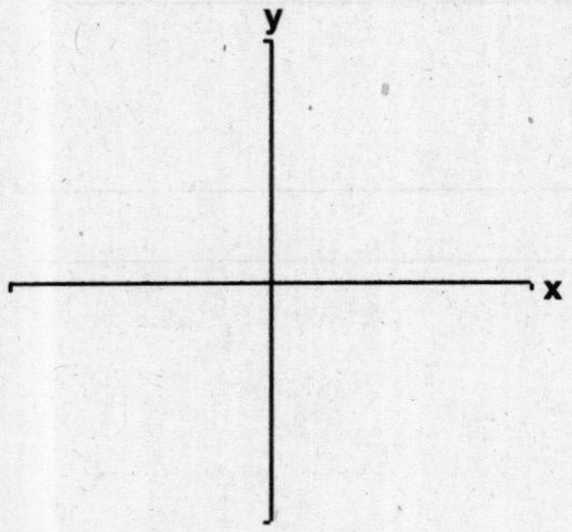

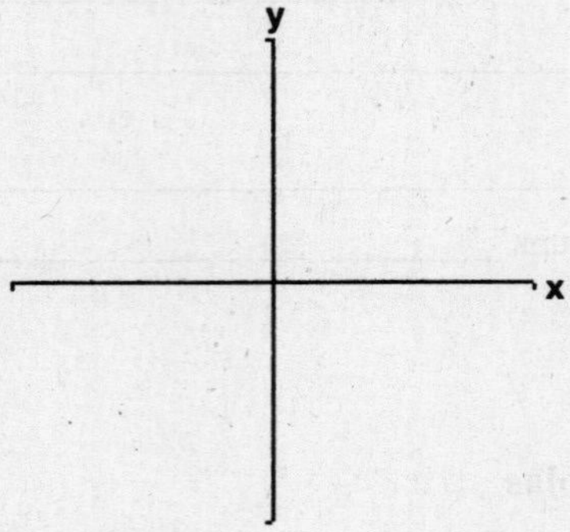

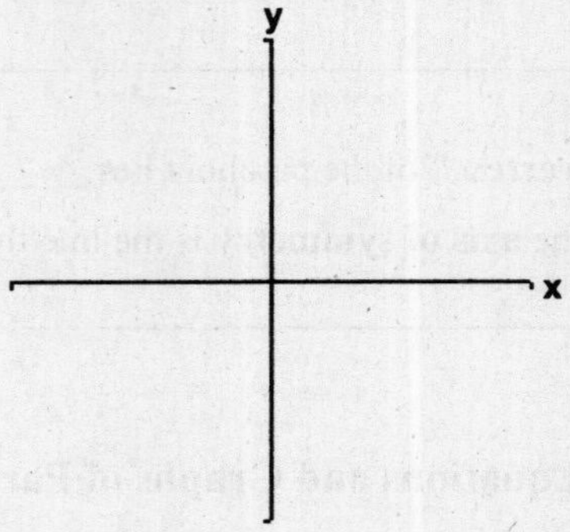

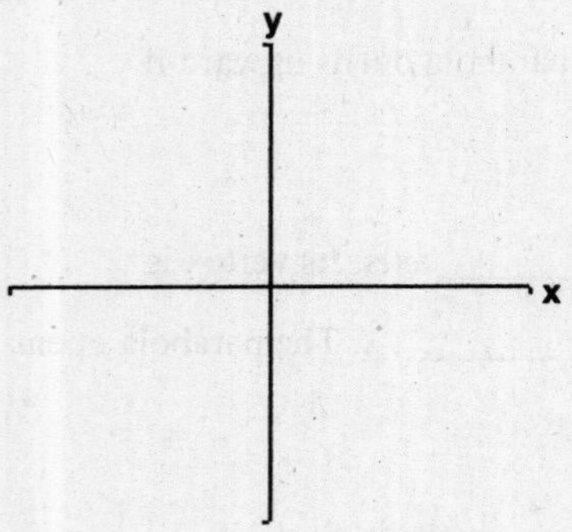

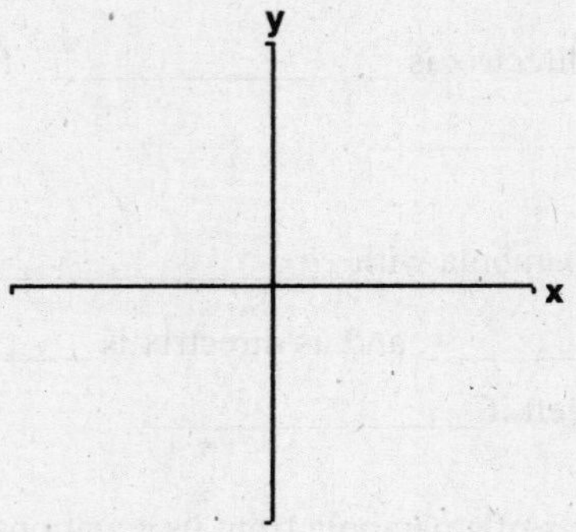

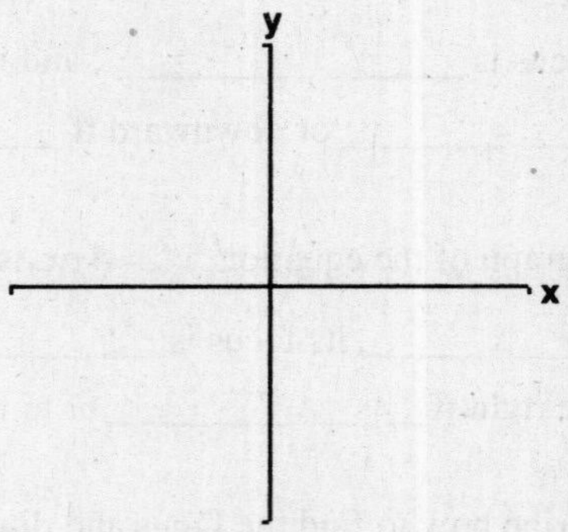

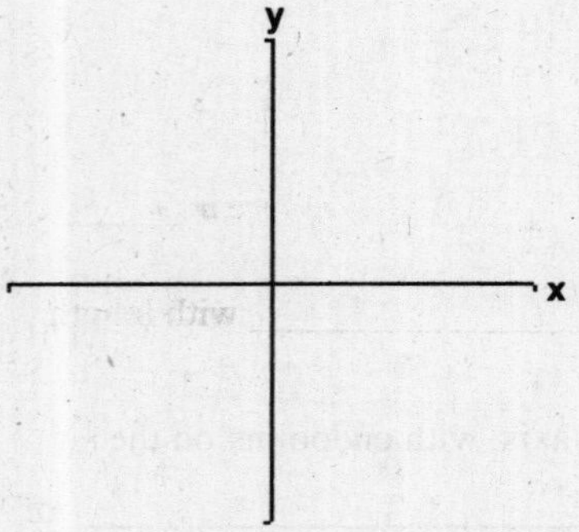

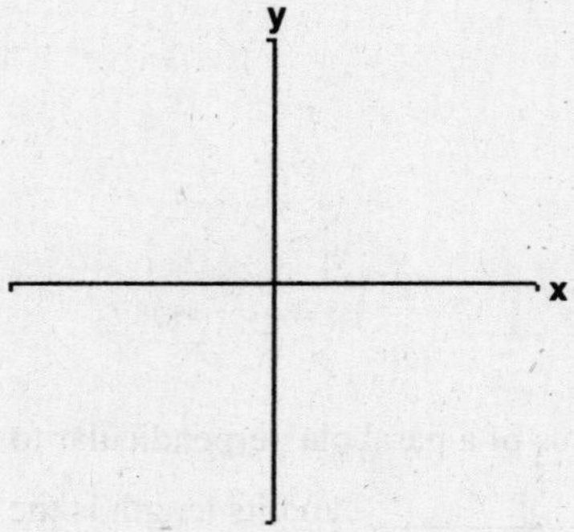

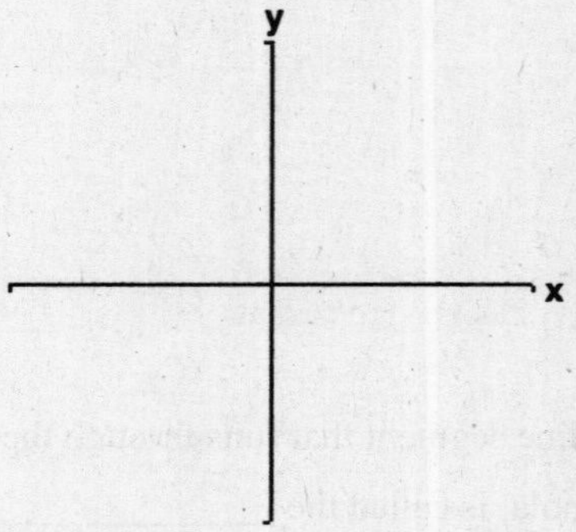

Homework Assignment

Page(s)

Exercises

Name _____ Date _____

11.2 Ellipses

I. Geometric Definition of an Ellipse

An **ellipse** is the set of points in the plane _____

_____. These two fixed points are the _____ of

the ellipse.

If the foci are on the x-axis, the ellipse crosses the x-axis at _____ and _____.

These points are called the _____ of the ellipse, and the segment that joins them is called

the _____. Its length is _____. The ellipse also crosses the y-axis at

_____ and _____. The segment that joins these points is called _____

_____, and it has length _____. The major axis is _____ the

minor axis. The origin is the _____ of the ellipse.

If the foci of the ellipse are placed on the y-axis rather than on the x-axis, then _____

_____, and we get a _____ ellipse.

II. Equations and Graphs of Ellipses

The graph of the equation $\dfrac{x^2}{a^2} + \dfrac{y^2}{b^2} = 1$ is an ellipse with center at the _____, where a _____ b.

Its vertices are located at _____. The major axis is _____ with length

_____. Its minor axis is _____ with length _____. The foci are located

at _____ and have the relationship _____.

The graph of the equation $\dfrac{x^2}{b^2} + \dfrac{y^2}{a^2} = 1$ is an ellipse with center at the _____, where a _____ b.

Its vertices are located at _____. The major axis is _____ with length

_____. Its minor axis is _____ with length _____. The foci are

located at _____ and have the relationship _____.

III. Eccentricity of an Ellipse

For the ellipse $\dfrac{x^2}{a^2} + \dfrac{y^2}{b^2} = 1$ or $\dfrac{x^2}{b^2} + \dfrac{y^2}{a^2} = 1$ (with $a > b > 0$), the **eccentricity** e is the number _____,

where $c = \sqrt{a^2 - b^2}$. The eccentricity of every ellipse satisfies _____.

If e is close to 1, then the ellipse is _____ in shape. If e is close to 0, then the ellipse is _____ in shape. The eccentricity is a measure of _____ _____ .

Ellipses have an interesting _____ that leads to a number of practical applications. If a light source is placed at one focus of a reflecting surface with elliptical cross sections, then _____ .

Give an example of a practical application of the reflection property of an ellipse and explain how it works.

Homework Assignment

Page(s)

Exercises

Name _____ Date _____

11.3 Hyperbolas

I. Geometric Definition of a Hyperbola

A **hyperbola** is the set of points in the plane _____

_____. These two fixed points are the

_____ of the hyperbola.

If the foci are on the x-axis, the hyperbola has x-intercepts at _____ and _____.

These points are called the _____ of the hyperbola.

If the hyperbola's foci are on the x-axis, does the graph of the hyperbola intersect the y-axis?

A hyperbola consists of two parts called its _____. The segment joining the two

vertices on the separate branches is the _____ of the hyperbola, and the

origin is called its _____.

If the foci of the hyperbola are placed on the y-axis rather than on the x-axis, this has the effect of _____

_____. This leads to a _____

_____.

II. Equations and Graphs of Hyperbolas

The graph of the equation $\dfrac{x^2}{a^2} - \dfrac{y^2}{b^2} = 1$, $a > 0$, $b > 0$, is a hyperbola with center at the _____. Its

vertices are located at _____. The transverse axis is _____ with

length _____. This hyperbola has asymptotes given by _____. The foci

are located at _____ and have the relationship _____.

The graph of the equation $\dfrac{y^2}{a^2} - \dfrac{x^2}{b^2} = 1$, $a > 0$, $b > 0$, is a hyperbola with center at the _____. Its

vertices are located at _____. The transverse axis is _____ with

length _____. This hyperbola has asymptotes given by _____. The foci

are located at _____ and have the relationship _____.

The asymptotes are lines that the hyperbola _____.

Asymptotes are an essential aid for graphing a hyperbola because they _____

_____. A convenient way to find the asymptotes for a hyperbola with horizontal

transverse axis is to first plot the points _____.
Then sketch horizontal and vertical segments through these points to construct a rectangle called the

_____ of the hyperbola. The slopes of the diagonals of the central box are $\pm b/a$,
so by extending them, we obtain _____.

List the steps for sketching a hyperbola.

Like parabolas and ellipses, hyperbolas have an interesting reflection property: light aimed at one focus of a
hyperbolic mirror is _____.

Give an example of how the hyperbola's reflection property is used in real life.

Homework Assignment

Page(s)

Exercises

Name _____ Date _____

11.4 Shifted Conics

I. Shifting Graphs of Equations

If h and k are positive real numbers, then replacing x by $x - h$ or $x + h$ and replacing y by $y - k$ or $y + k$ has the following effect(s) on the graph of any equation in x and y.

Replacement	How the graph is shifted
1. x replaced by $x - h$	_____
2. x replaced by $x + h$	_____
3. y replaced by $y - k$	_____
4. y replace by $y + k$	_____

II. Shifted Ellipses

If we shift an ellipse so that its center is at the point (h, k), instead of at the origin, then its equation becomes

_____

III. Shifted Parabolas

If we shift a parabola so that its center is at the point (h, k), instead of at the origin, then its equation becomes

_____ for a parabola with a vertical axis or _____ for a parabola with a vertical axis.

IV. Shifted Hyperbolas

If we shift a hyperbola so that its center is at the point (h, k), instead of at the origin, then its equation becomes

_____ for a hyperbola with a horizontal transverse axis or

_____ for a hyperbola with a vertical transverse axis.

V. The General Equation of a Shifted Conic

If we expand and simplify the equations of any of the shifted conics, then we will always obtain an equation of

the form _____, where A and C are not both 0. Conversely,

if we begin with an equation of this form, then we can _____ to

see which type of conic section the equation represents.

In some cases the graph of the equation turns out to be just a pair of lines or a single point, or there may be no

graph at all. These cases are called _____.

The graph of the equation $Ax^2 + Cy^2 + Dx + Ey + F = 0$, where A and C are not both 0, is a conic or _____

_____. In the nondegenerate cases the graph is

1. a(n) _____ if A or C is 0,

2. a(n) _____ if A and C have the same sign (or a circle if _____),

3. a(n) _____ if A and C have opposite signs.

Homework Assignment

Page(s)

Exercises

Name _____ Date _____

11.5 Rotation of Axes

I. Rotation of Axes

If the x- and y-axes are rotated through an acute angle ϕ about the origin to produce a new pair of axes, these

new axes are called the _____. A point P that has coordinates

(x, y) in the old system has coordinates _____ in the new system.

Suppose the x- and y-axes in a coordinate plane are rotated through the acute angle ϕ to produce the X- and
Y-axes. Then the coordinates (x, y) and (X, Y) of a point in the xy- and XY-planes are related as follows:

$x = $ _____ $X = $ _____

$y = $ _____ $Y = $ _____

If the coordinate axes are rotated through an angle of 45°, describe how to find the XY-coordinates of the point
with xy-coordinates (1, 5).

II. General Equation of a Conic

To eliminate the xy-term in the general conic equation $Ax^2 + Bxy + Cy^2 + Dx + Ey + F = 0$, rotate the axes
through the acute angle ϕ that satisfies _____.

III. The Discriminant

The graph of the equation $Ax^2 + Bxy + Cy^2 + Dx + Ey + F = 0$ is either a conic or a degenerate conic. In the
nondegenerate cases, the graph is

1. a(n) _____ if $B^2 - 4AC = 0$,

2. a(n) _____ if $B^2 - 4AC < 0$;

3. a(n) _____ if $B^2 - 4AC > 0$.

The quantity $B^2 - 4AC$ is called the _____ of the equation.

The discriminant is unchanged by any rotation and, thus, is said to be _____.

Additional notes

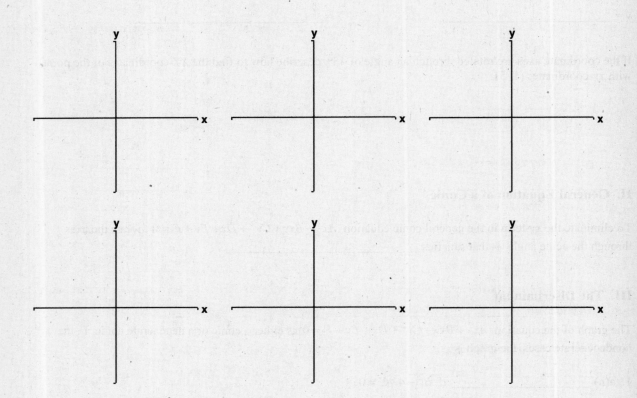

Homework Assignment

Page(s)

Exercises

Name _____ Date _____

11.6 Polar Equations of Conics

I. A Unified Geometric Description of Conics

Let F be a fixed point (the _____), ℓ a fixed line (the _____), and let e be a fixed positive number (the _____). The set of all points P such that the ratio of the distance from P to F to the distance from P to ℓ is the constant e is a _____.

That is, the set of all points P such that _____ is a conic. The conic is a parabola if _____, an ellipse if _____, or a hyperbola if _____.

II. Polar Equations of Conics

A polar equation of the form $r = \dfrac{ed}{1 \pm e\cos\theta}$ or $r = \dfrac{ed}{1 \pm e\sin\theta}$ represents a conic with one focus at the origin and with eccentricity e. The conic is:

1. a parabola if _____,

2. an ellipse if _____,

3. a hyperbola if _____.

To graph the polar equation of a conic, we first _____
_____. For a parabola, the _____ is perpendicular to the directrix. For an ellipse, the _____ is perpendicular to the directrix. For a hyperbola, the _____ is perpendicular to the directrix.

To graph a polar conic, it is helpful to plot the points for which $\theta =$ _____. Using these points and a knowledge of the type of conic (which we obtain from the eccentricity), we can easily get a rough idea of _____.

When we rotate conic sections, it is much more convenient to use polar equations than _____. We use the fact that the graph of $r = f(\theta - \alpha)$ is the graph of _____
_____.

When e is close to 0, an ellipse is _____, and it becomes more elongated as _____. When $e = 1$, the conic is _____. As e increases beyond 1, the conic is _____.

Additional notes

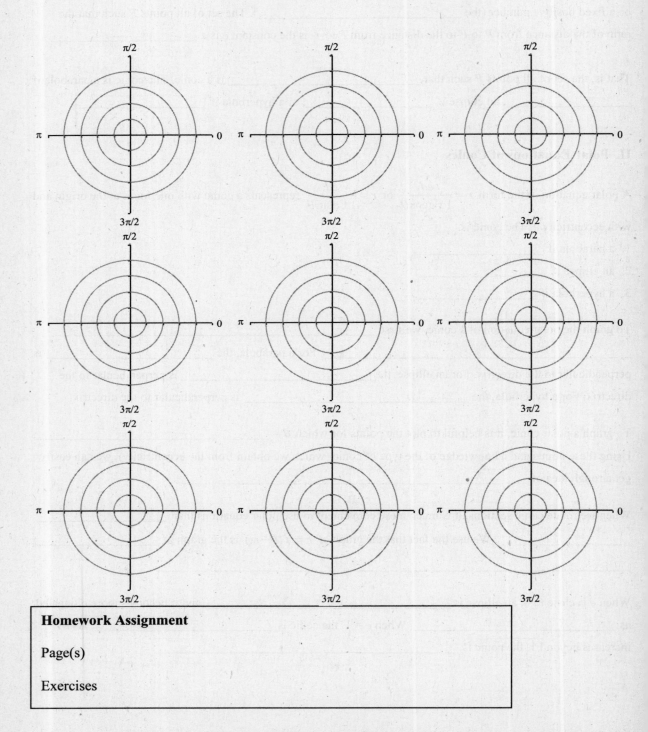

Homework Assignment

Page(s)

Exercises

Chapter 12   Sequences and Series

12.1   Sequences and Summation Notation

I. Sequences

A *sequence* is a set of numbers written in a _____.

A **sequence** is _____.

The values $f(1), f(2), f(3), \ldots$ are called the _____ of the sequence.

To specify a procedure for finding all the terms of a sequence, _____
_____.

The presence of $(-1)^n$ in the sequence has the effect of _____
_____.

II. Recursively Defined Sequences

A **recursive** sequence is a sequence in which _____
_____.

The **Fibonacci sequence**, given as _____ was
named after the 13th century Italian mathematician who used it to solve a problem about the breeding of rabbits.

III. The Partial Sums of a Sequence

For the sequence $a_1, a_2, a_3, a_4, \ldots, a_n, \ldots$ the **partial sums** are

$$S_1 = \text{_____}$$

$$S_2 = \text{_____}$$

$$S_3 = \text{_____}$$

$$S_4 = \text{_____}$$

$$\vdots$$

$$S_n = \text{_____}$$

S_1 is called the _____. S_2 is the _____, and

so on. S_n is called the _____. The sequence $S_1, S_2, S_3, \ldots, S_n, \ldots$ is

called the _____.

IV. Sigma Notation

Given a sequence $a_1, a_2, a_3, a_4, \ldots$, we can write the sum of the first n terms using _____

_____, which derives its name from the Greek letter _____.

The notation is used as follows: $\displaystyle\sum_{k=1}^{n} a_k =$ _____

The left side of this expression is read as _____ _____. The letter k is

called the _____, or the _____, and the

idea is to replace k in the expression after the sigma by _____,

and add the resulting expressions.

Let $a_1, a_2, a_3, a_4, \ldots$ and $b_1, b_2, b_3, b_4, \ldots$ be sequences. Then for every positive integer n and any real number c, complete each of the following properties of sums.

1. $\displaystyle\sum_{k=1}^{n} (a_k + b_k) =$

    _____

2. $\displaystyle\sum_{k=1}^{n} (a_k - b_k) =$

    _____

3. $\displaystyle\sum_{k=1}^{n} c a_k =$

    _____

Homework Assignment

Page(s)

Exercises

Name _____ Date _____

12.2   Arithmetic Sequences

I. Arithmetic Sequences

An **arithmetic sequence** is a sequence of the form _____.

The number a is the _____, and d is the _____ of the sequence.

The **nth term** of an arithmetic sequence is given by _____.

The number d is called the common difference because _____
_____.

An arithmetic sequence is determined completely by _____
_____. Thus, if we know the first two terms of an arithmetic sequence, then _____
_____.

II. Partial Sums of Arithmetic Sequences

For the arithmetic sequence $a_n = a + (n-1)d$ the **nth partial sum**
$S_n = a + (a+d) + (a+2d) + (a+3d) + \cdots + [a+(n-1)d]$ is given by either of the following formulas.

1.
_____

2.
_____

Give an example of a real-life situation in which a partial sum of an arithmetic sequence is used.

Additional notes

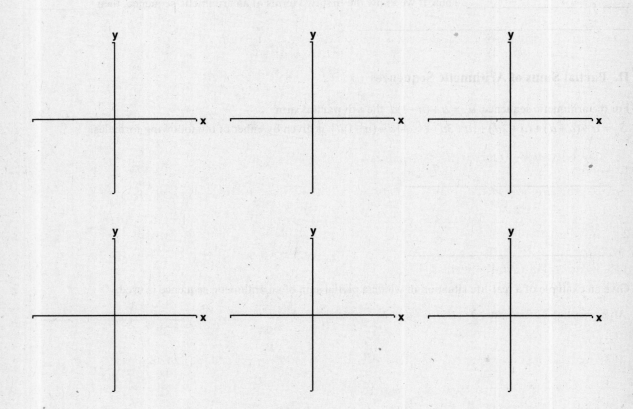

Homework Assignment

Page(s)

Exercises

Name _____ Date _____

12.3 Geometric Sequences

I. Geometric Sequences

A *geometric* sequence is generated when we start with a number a and _____

_____.

An **geometric sequence** is a sequence of the form _____. The number a is the _____

_____, and r is the _____ of the sequence. The **nth term** of a geometric sequence

is given by _____.

The number r is called the common ratio because _____

_____.

Give a real-life example of a geometric sequence.

II. Partial Sums of Geometric Sequences

For the geometric sequence $a_n = ar^{n-1}$ the **nth partial sum** $S_n = a + ar + ar^2 + ar^3 + ar^4 + \cdots + ar^{n-1}$ $(r \neq 1)$

is given by _____.

III. What Is an Infinite Series?

An expression of the form $\displaystyle\sum_{k=1}^{\infty} a_k = a_1 + a_2 + a_3 + a_4 + \cdots$ is called an _____.

If the partial sum S_n gets close to a finite number S as n gets large, we say that the infinite series _____

_____. The number S is called the _____.

If an infinite series does not converge, we say that the series _____.

IV. Infinite Geometric Series

An **infinite geometric series** is a series of the form _____.

If $|r| < 1$, then the infinite geometric series $\displaystyle\sum_{k=1}^{\infty} ar^{k-1} = a + ar + ar^2 + ar^3 + \cdots$ _____ and

has the sum _____ .

If $|r| \geq 1$, then the series _____ .

Additional notes

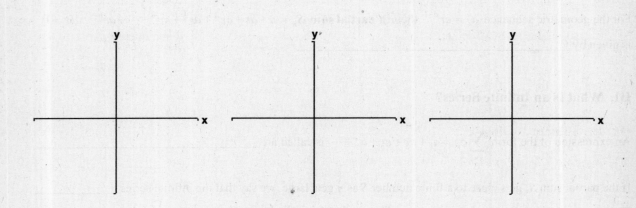

Homework Assignment

Page(s)

Exercises

Name _____ Date _____

12.4 Mathematics of Finance

I. The Amount of an Annuity

An **annuity** is _____.

The payments are usually made at _____. The **amount**

of an annuity is _____

_____.

In general, the regular annuity payment is called the _____ and is denoted by

R. We also let i denote the _____ and let n denote _____

_____. We always assume that the time period in which interest is compounded is equal to

the _____. The amount A_f of an annuity consisting of n

regular equal payments of size R with interest rate i per time period is given by _____.

II. The Present Value of an Annuity

The **present value of an annuity** is the amount A_p that _____

_____.

The **present value** A_p of an annuity consisting of n regular equal payments of size R and interest rate i per time

period is given by _____.

III. Installment Buying

When you buy a house or a car by installment, the payments that you make are _____

_____.

For an installment buying situation, if a loan A_p is to be repaid in n regular equal payments with interest rate i

per time period, then the size R of each payment is given by _____.

Additional notes

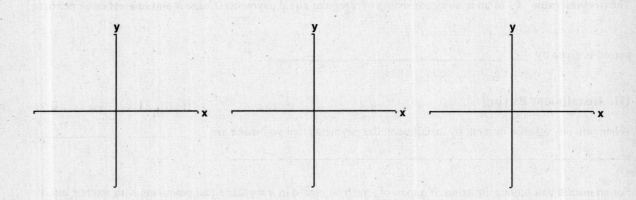

Homework Assignment	
Page(s)	
Exercises	

Name _____ Date _____

12.5 Mathematical Induction

I. Conjecture and Proof

What is a conjecture?

A mathematical **proof** is _____
_____.

II. Mathematical Induction

In mathematical induction, the induction step leads us _____
_____.

The **Principle of Mathematical Induction** states that for each natural number n, let $P(n)$ be a statement depending on n. Suppose that the following two conditions are satisfied.

1. _____.

2. _____.

Then $P(n)$ is true for _____.

Describe how to apply the Principle of Mathematical Induction.

Notice that we do not prove that $P(k)$ is _____. We only show that *if* $P(k)$ is true, then
_____. The assumption that $P(k)$ is true is
called the _____.

Complete each of the following formulas for the sums of powers.

0. $\displaystyle\sum_{k=1}^{n} 1 =$ _____

1. $\displaystyle\sum_{k=1}^{n} k =$ _____

2. $\displaystyle\sum_{k=1}^{n} k^2 =$ _____

3. $\displaystyle\sum_{k=1}^{n} k^3 =$ _____

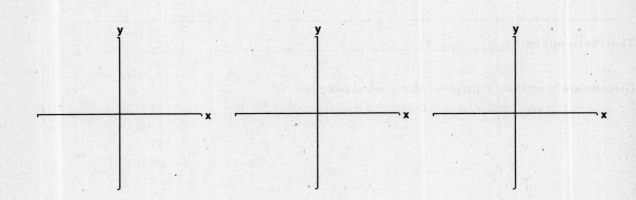

Homework Assignment

Page(s)

Exercises

Name _____ Date _____

12.6 The Binomial Theorem

An expression of the form $a+b$ is called a _____.

I. Expanding $(a+b)^n$

List some of the simple patterns that emerge from the expansion of $(a+b)^n$.

Write the first nine rows of Pascal's Triangle.

The key property of Pascal's Triangle is that every entry (other than a 1) is _____

_____.

Describe how to use Pascal's Triangle to expand a binomial $(a+b)^n$.

II. The Binomial Coefficients

Although Pascal's Triangle is useful in finding the binomial expansion for reasonably small values of n, it isn't

practical for finding $(a+b)^n$ for large values of n because _____

_____

_____.

The product of the first n natural numbers is denoted by _____ and is called _____.

$n! =$ _____

$0! =$ _____

Let n and r be nonnegative integers with $r \leq n$. The **binomial coefficient** is denoted by $\binom{n}{r}$ and is defined by

_____.

The binomial coefficient $\binom{n}{r}$ is always a _____ number. Also, $\binom{n}{r}$ is equal to _____.

The key property of the binomial coefficients is that for any nonnegative integers r and k with $r \leq k$,

_____.

III. The Binomial Theorem

The Binomial Theorem states that $(a+b)^n =$ _____.

The general term that contains a^r in the expansion of $(a+b)^n$ is _____.

Homework Assignment

Page(s)

Exercises

Chapter 13   Limits: A Preview of Calculus

13.1   Finding Limits Numerically and Graphically

I. Definition of Limit

We write $\lim_{x \to a} f(x) = L$ and say "the limit of $f(x)$, as x approaches a, equals L" if _____

_____

_____.

This says that the values of $f(x)$ get closer and closer to _____

_____.

II. Estimating Limits Numerically and Graphically

Describe how to estimate a limit numerically.

Describe how to estimate a limit graphically.

III. Limits That Fail to Exist

Functions do not necessarily approach a finite value at every point. In other words, it's possible for _____

_____.

Describe three situations in which a limit may fail to exist.

To indicate that a function with a vertical asymptote has a limit that fails to exist, we use the notation

$\lim\limits_{x \to a} f(x) = \infty$, which expresses the particular way in which the limit does not exist: $f(x)$ can be made as large

as we like by _____.

IV. One-Sided Limits

We write $\lim\limits_{x \to a^-} f(x) = L$ and say _____

_____ if we can

make the values of $f(x)$ arbitrarily close to L by taking x to be sufficiently close to a and x less than a.

Similarly, if we require that x be greater than a, we get _____

_____, and we write $\lim\limits_{x \to a^+} f(x) = L$.

By comparing the definitions of two-sided and one-sided limits, we see that the following is true: $\lim\limits_{x \to a} f(x) = L$

if and only if _____. Thus if the left-hand and right-hand

limits are different, the two-sided limit _____.

y | x

y | x

y | x

Homework Assignment

Page(s)

Exercises

Name _____ Date _____

13.2 Finding Limits Algebraically

I. Limit Laws

Suppose that c is a constant and that the following limits exist: $\lim\limits_{x \to a} f(x)$ and $\lim\limits_{x \to a} g(x)$, then give each of the following limits.

1. Limit of a Sum: _____

2. Limit of a Difference: _____

3. Limit of a Constant Multiple: _____

4. Limit of a Product: _____

5. Limit of a Quotient: _____

6. Limit of a Power: _____

7. Limit of a Root: _____

State each of these five laws verbally.

II. Applying the Limit Laws

Complete each of the following special limits.

1. $\lim\limits_{x \to a} c =$ _____

2. $\lim\limits_{x \to a} x =$ _____

3. $\lim\limits_{x \to a} x^n =$ _____

4. $\lim_{x \to a} \sqrt[n]{x} = $ _____

If f is a polynomial or a rational function and a is in the domain of f, then the limit of f may be found by direct

substitution, that is, $\lim_{x \to a} f(x) = $ _____. Functions with this direct

substitution property are called _____.

III. Finding Limits Using Algebra and the Limit Laws

Example 1   Evaluate $\lim_{y \to 4} \dfrac{16 - 8y + y^2}{y - 4}$.

IV. Using Left- and Right-Hand Limits

A two-sided limit exists if and only if _____.

When computing one-sided limits, use the fact that _____

_____.

Example 2   Let $f(x) = \begin{cases} x^2 - 3x & \text{if } x \geq 3 \\ x & \text{if } x < 3 \end{cases}$. Determine whether $\lim_{x \to 3} f(x)$ exists.

Example 3   Let $f(x) = \begin{cases} x^2 - 3x & \text{if } x \geq 0 \\ x & \text{if } x < 0 \end{cases}$. Determine whether $\lim_{x \to 0} f(x)$ exists.

Homework Assignment

Page(s)

Exercises

Name _____ Date _____

13.3 Tangent Lines and Derivatives

I. The Tangent Problem

A *tangent line* is a line that _____.

We sometimes refer to the slope of the tangent line to a curve at a point as the _____

_____. The idea is that if we zoom in far enough toward the point, the curve

looks _____.

The **tangent line** to the curve $y = f(x)$ at the point $P(a, f(a))$ is the line through P with slope

$m = $ _____, provided that this limit exists.

Another expression for the slope of a tangent line is $m = $ _____.

II. Derivatives

The **derivative of a function f at a number a,** denoted by $f'(a)$, is _____ if

this limit exists.

We see from the definition of a derivative that the number $f'(a)$ is the same as _____

_____.

III. Instantaneous Rates of Change

If $y = f(x)$, the **instantaneous rate of change of y with respect to x** at $x = a$ is the limit of the average rates of

change as x approaches a: instantaneous rate of change = _____.

List two different ways of interpreting the derivative.

In the special case in which $x = t$ = time and $s = f(t)$ = displacement (directed distance) at time t of an object traveling in a straight line, the instantaneous rate of change is called the _____.

Additional notes

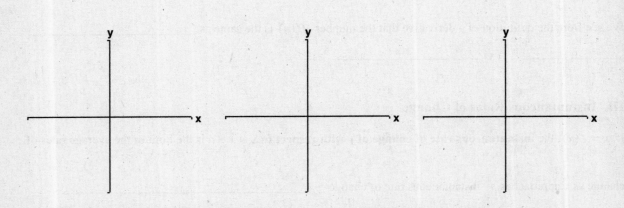

Homework Assignment

Page(s)

Exercises

Name _____ Date _____

13.4 Limits at Infinity; Limits of Sequences

I. Limits at Infinity

We use the notation $\lim\limits_{x \to \infty} f(x) = L$ to indicate that the values of $f(x)$ become _____

_____ .

Let f be a function defined on some interval (a, ∞). Then by the definition of **limit at infinity** $\lim\limits_{x \to \infty} f(x) = L$

means that the values of $f(x)$ can be made _____ .

List various ways of reading the expression $\lim\limits_{x \to \infty} f(x) = L$.

Let f be a function defined on some interval $(-\infty, a)$. Then by the definition of a **limit at negative infinity**

$\lim\limits_{x \to -\infty} f(x) = L$ means that the values of $f(x)$ can be made _____

_____ .

How can one read the expression $\lim\limits_{x \to -\infty} f(x) = L$?

The line $y = L$ is called _____ of the curve $y = f(x)$ if either

$\lim\limits_{x \to \infty} f(x) = L$ or $\lim\limits_{x \to -\infty} f(x) = L$.

The Limit Laws studied earlier in this chapter _____ for limits at infinity.

If k is any positive integer, then $\lim\limits_{x \to \infty} \dfrac{1}{x^k} =$ _____ and $\lim\limits_{x \to -\infty} \dfrac{1}{x^k} =$ _____ .

II. Limits of Sequences

A sequence $a_1, a_2, a_3, a_4, \ldots$ has the **limit** L and we write $\lim\limits_{n \to \infty} a_n = L$ or $a_n \to L$ as $n \to \infty$ if the nth term a_n of

the sequence can be made _____ by taking n sufficiently large. If $\lim\limits_{n \to \infty} a_n$

exists, we say the sequence _____. Otherwise, we say the

sequence _____.

If $\lim\limits_{x \to \infty} f(x) = L$ and $f(n) = a_n$ when n is an integer, then $\lim\limits_{n \to \infty} a_n =$ _____.

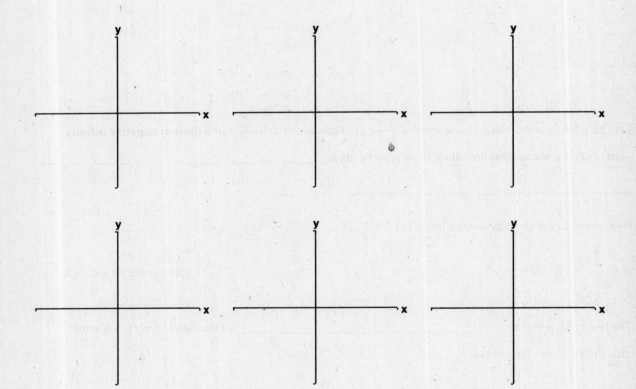

Homework Assignment

Page(s)

Exercises

Name _____ Date _____

13.5 Areas

I. The Area Problem

Describe the *area problem* found in calculus.

Describe the approach taken in calculus to finding the area of a region S with curved sides.

II. Definition of Area

The area A of the region S that lies under the graph of the continuous function f is the limit of the sum of the areas of approximating rectangles:

_____

Using sigma notation, we write this as follows:

                                        _____

In using this formula for area, remember that Δx is the _____ of an approximating rectangle, x_k

is the _____ of the kth rectangle, and $f(x_k)$ is its _____.

Complete each of the following formulas.

$\Delta x = $ _____

$x_k = $ _____

$f(x_k) = $ _____

Additional notes

Chapter 14   Probability and Statistics

14.1   Counting

I. The Fundamental Counting Principle

Suppose that two events occur in order. The Fundamental Counting Principle states that if the first can occur in *m* ways and the second can occur in *n* ways (after the first has occurred), then the two events can occur in order in _____ ways.

Explain how the Fundamental Counting Principle may be extended to any number of events.

II. Counting Permutations

A **permutation** of a set of distinct objects is _____.

The number of permutations of *n* objects is _____.

In general, if a set has *n* elements, then the number of ways of ordering *r* elements from the set is denoted by *P(n, r)* and is called the _____.

The number of permutations of *n* objects taken *r* at a time is given by _____.

III. Counting Combinations

When counting combinations, order _____ important.

A **combination** of *r* elements of a set is _____
_____. If the set has *n* elements, then the number of combinations of *r* elements is denoted by *C(n, r)* and is called the _____.

The number of combinations of *n* objects taken *r* at a time is given by _____.

The key difference between permutations and combinations is _____.

IV. Problem Solving with Permutations and Combinations

List the guidelines for solving counting problems.


```
┌─────────────────────────────────────────────────────────────────────────┐
│ Homework Assignment                                                       │
│                                                                           │
│ Page(s)                                                                   │
│                                                                           │
│ Exercises                                                                 │
│                                                                           │
│                                                                           │
└─────────────────────────────────────────────────────────────────────────┘
```

Name _____ Date _____

14.2 Probability

I. What is Probability?

An _____ is a process, such as tossing a coin, that gives definite results, called the

_____ of the experiment. The **sample space** of an experiment is _____

_____.

We will be concerned only with experiments for which all the outcomes are _____.

If S is the sample space of an experiment, then an **event** E is _____.

Let S be the sample space of an experiment in which all outcomes are equally likely, and let E be an event. Then

the **probability** of E, written $P(E)$, is _____.

The probability $P(E)$ of an event is a number between _____. The closer the

probability of an event is to 1, the _____ the event is to happen; the closer the

probability of an event is to 0, the _____ the event is to happen. If $P(E) = 0$, then E is

called the _____.

II. Calculating Probability by Counting

To find the probability of an event, we do not need to list all the elements _____

_____. We only need _____ in these sets.

The _____ learned earlier will be very useful here.

III. The Complement of an Event

The **complement** of an event E is the set of outcomes in the sample space that is not in E. We denote the

complement of E by _____.

Let S be the sample space of an experiment and let E be an event. Then the probability of E', the complement of

E, is _____.

IV. The Union of Events

If E and F are events in a sample space S, then the probability of E or F, that is the union of these events, is

_____.

Two events that have no outcome in common are said to be _____. In other words, the events E and F are mutually exclusive if _____. So if the events E and F are mutually exclusive, then $P(E \cap F) = $ _____.

If events E and F are mutually exclusive then the probability of the union of these two mutually exclusive events is _____.

V. Conditional Probability and the Intersection of Events

Let E and F be events in a sample space S. The **conditional probability of E given that F occurs** is

$P(E \mid F) = $ _____.

If E and F are events in a sample space S, then the probability of E *and* F, that is, the intersection of these two events, is $P(E \cap F) = $ _____.

When the occurrence of one event does not affect the probability of the occurrence of another event, we say that the events are _____. This means that the events E and F are independent if _____ and _____.

Homework Assignment

Page(s)

Exercises

Name _____ Date _____

14.3 Binomial Probability

I. Binomial Probability

A **binomial experiment** is one in which there are _____ outcomes, which are called "success" and "failure."

Binomial Probability

An experiment has two possible outcomes called "success" and "failure," with $P(\text{success}) = p$ and

$P(\text{failure}) =$ _____. The probability of getting exactly r successes in n independent trials of the

experiment is $P(r \text{ successes in } n \text{ trials}) =$ _____.

II. The Binomial Distribution

The function that assigns to each outcome its corresponding probability is called a _____

_____. A bar graph of a probability distribution in which the width of each bar is 1 is

called a _____.

A probability distribution in which all outcomes have the same probability is called a(n) _____

_____. The probability distribution of a binomial experiment is called a(n) _____

_____.

The sum of the probabilities in a probability distribution is _____, because the sum is the

probability of the occurrence of any outcome in the sample space.

Additional notes

Homework Assignment

Page(s)

Exercises

Name _____ Date _____

14.4 Expected Value

I. Expected Value

A game gives payouts $a_1, a_2, \ldots, a_n$ with probabilities $p_1, p_2, \ldots, p_n$. The **expected value** (or **expectation**) E of

this game is _____.

Give an example of a real-life application of expected value.

II. What is a Fair Game?

A **fair game** is _____. So if you play a fair game many

times you would expect, on average, to _____.

Describe the use of fair games in casinos.

Invent a simple fair game and show that it is fair.

Additional notes

Homework Assignment

Page(s)

Exercises

Name _____ Date _____

14.5 Descriptive Statistics (Numerical)

The things that a data set describes are called _____. The property of the individuals that is described by the data is called a _____. We will be studying **one-variable data,** where _____.

The first goal of statistics is _____. A **summary statistic** is _____.

I. Measures of Central Tendency: Mean, Median, Mode

One way to make sense of data is to find _____ or the _____ of the data. Any such number is called a _____. One such measure is the _____.

Let $x_1, x_2, \ldots, x_n$ be n data points. The **mean** (or **average**), denoted by $\bar{x}$, is the sum divided by n:

_____

Another measure of central tendency is the _____, which is the middle number of an ordered list of numbers.

Let $x_1, x_2, \ldots, x_n$ be n data points, written in increasing order. If n is odd, the **median** is _____ _____. If n is even, the **median** is _____.

If a data set includes a number far away from the rest of the data, that data point is called _____.

If a data set has outliers, which is a better indicator of the central tendency of the data, the median or the mean?

_____

The **mode** of a data set is _____.

The mode has the advantage of _____.

A data set with two modes is called _____. A data set such as 3, 5, 7, 9, 11 has _____ mode.

II. Organizing Data: Frequency Tables and Stemplots

A **frequency table** for a set of data is _____
_____. The _____ is most easily determined from a frequency table.

A **stemplot** (or **stem-and-leaf plot**) organizes data by _____

_____. Each number in the data is written as a **stem** consisting of _____

and a **leaf** consisting of _____. Numbers with the same stem are _____

_____, with the stem written _____.

III. Measures of Spread: Variance and Standard Deviation

Measures of spread (also called **measures of dispersion**) describe _____

_____.

Let $x_1, x_2, \ldots, x_N$ be N data points and let $\bar{x}$ be their mean. The **standard deviation** of the data is

_____. The **variance** is _____, the square of the standard

deviation.

IV. The Five Number Summary: Box Plots

A simple indicator of the spread of data is the location of _____.

Other indicators of spread are the _____. The median divides a data set _____.

The median of the lower half of the data is called _____. The median of the

upper half of the data is called _____. Together these values give a

good picture of _____.

The **five-number summary** for a data set are the five numbers below, written in the indicated order.

_____

A simple indicator of spread is the **range,** which is _____

_____. This can be compared to the spread of the middle of the data as

measured by the **interquartile range (IQR),** which is _____

_____.

A **box plot** (also called a **box-and-whisker plot**) is a method for _____

_____. The plot consists of _____

_____. The box is divided by a line segment at _____.

The **whiskers** are _____

_____.

When working with quartiles, the median of the data is also called _____.

Name _____ Date _____

14.6 Descriptive Statistics (Graphical)

I. Data in Categories

Numerical data is _____.

Categorical data is _____
_____. Categorical data can be represented by _____.

A **bar graph** consists of _____. The height of
each bar is proportional to _____.
So the y-axis has a _____ corresponding to the number or the proportion of the
individuals in each category. The labels on the x-axis describe _____.

A **pie chart** consists of _____. The
central angle of each sector is proportional to _____.
Each sector is labeled with _____.

II. Histograms and the Distribution of Data

To visualize the distribution of one-variable numerical data, we first _____
_____. To do so, we divide the range of the data into _____
_____, called **bins**. To draw a histogram of the data we first label
the bins on the x-axis, and then _____; the height (and hence
also the area) of each rectangle is proportional to _____.

A histogram gives a visual representation of how the data are _____ in the different bins. The
histogram allows us to determine if the data are _____ about the mean. If the histogram
has a long "tail" on the right, we say the data are _____. Similarly, if there is a
long tail on the left, the data are _____. Since the area of each bar in the
histogram is proportional to the number of data points in that category, it follows that the median of the data is
located at _____.

III. The Normal Distribution

Most real-world data are distributed in a special way called a _____.

The **standard normal distribution** (or **standard normal curve**) is modeled by the function

_____.

This distribution has mean _____ and standard deviation _____. **Normal distributions** with different means and standard deviations are modeled by transformations (shifting and stretching) of the above function. Specifically, the normal distribution with mean μ and standard deviation σ

is modeled by the function _____.

All normal distributions have the same general shape, called a _____.

For normally distributed data with mean μ and standard deviation σ, we have the following facts, called the **Empirical Rule.**

- Approximately _____ of the data are between $\mu - \sigma$ and $\mu + \sigma$

- Approximately _____ of the data are between $\mu - 2\sigma$ and $\mu + 2\sigma$

- Approximately _____ of the data are between $\mu - 3\sigma$ and $\mu + 3\sigma$

Homework Assignment

Page(s)

Exercises

Name _____ Date _____

14.7   Introduction to Statistical Thinking

In statistical thinking we make judgments about an entire population based on _____. A **population** consists of _____. A **sample** is _____

_____.

In **survey sampling,** data are collect through _____. In **observational studies,** data are collected by _____. In **experimental studies,** data are collected by _____.

I. The Key Role of Randomness

A **random sample** is one that is selected _____. In statistical thinking, we expect a random sample to _____ as the population. The larger the sample size, the more closely the sample properties _____

_____.

In general, **non-random samples** are generated when there is _____. Such samples are useless for statistical purposes because _____

_____.

A **simple random sample** is one in which _____

_____. To satisfy this requirement the sample method used must be _____

_____ with respect to the property being measured.

List four common types of sampling bias.

Many of these biases are a result of **convenience sampling,** in which individuals are sampled _____

_____.

II. Design of Experiments

In observational studies the researcher has no control over the factors affecting the property being studied. Extraneous or unintended variables that systematically affect the property being studied are called _____ _____. Such variables are said to **confound,** or _____, the results of the study.

To eliminate or vastly reduce the effects of confounding variables, researchers often conduct experiments so that such variables can be _____. In an experimental study, two groups are selected—a **treatment group** (in which _____) and a **control group** (in which _____). The individuals in the experiment are called _____. The goal is to measure the **response** of the subjects to the treatment—that is, _____. The next step is to make sure that the two groups are as similar as possible, except for _____. If the two groups are alike except for the treatment, then any statistical difference in response between the groups can be confidently attributed to _____.

A common confounding factor is the _____, in which patients who think they are receiving a medication report an improvement even though the "treatment" they received was a **placebo,** _____ _____.

III. Margin of Error and Sample Size

Statistical conclusions are based on probability and are always accompanied by a _____. The 95% confidence level means that there is less than a 5% chance (or 0.05 probability) that the result obtained from the sample _____. In the popular press, poll results are accompanied by _____.

At the 95% confidence level, the margin of error d and the sample size n are related by the formula

_____.

IV. Two-Variable Data and Correlation

Two-variable data measures _____. Two-variable data can be graphed in a coordinate plane resulting in a _____. Analyzing such data mathematically by finding the line that best fits the data is called finding the _____. Associated with the regression line is a _____, which is a measure of how well the data fit along the regression line, or how well the two variables are correlated. The correlation coefficient r is between _____ and _____. If r is close to zero, the variables have _____. The closer r is to 1 or −1 , the closer _____

_____.

In statistics, a question of interest when studying two-variable data is whether or not the correlation is statistically significant. That is, what is the probability that the correlation in the sample is due to _____

_____? If the sample consists of only three individuals, even a strong correlation coefficient _____

_____. On the other hand, for a large sample a small correlation coefficient may be significant. This is because if there is no correlation at all in the population, it's very unlikely that a large random sample would produce data that have a linear trend, whereas a small sample is more likely to produce correlated data by _____.

Correlation is _____ causation.

Additional notes

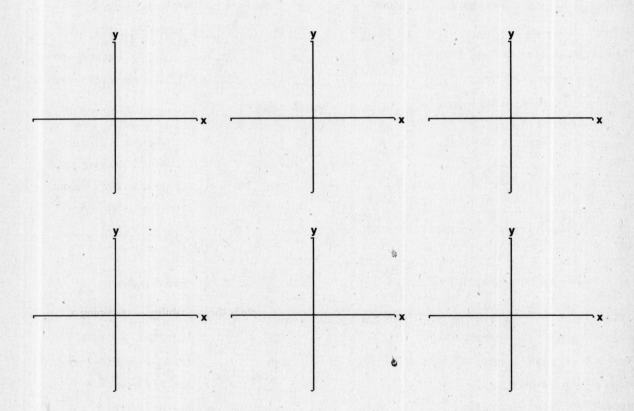

Homework Assignment

Page(s)

Exercises

Name _____ Date _____

14.8 Introduction to Inferential Statistics

The goal of **inferential statistics** is to infer information about an entire population by _____

_____ .

I. Testing a Claim about a Population Proportion Intuitively

In statistics a **hypothesis** is a statement or claim about _____ .

In this section we study hypotheses about the true proportion p of individuals in the population with _____

_____ .

Suppose that individuals having a particular property are assumed to form a proportion p_0 of a population. To

answer the question of how this assumption compares to the true proportion p of these individuals, we begin by

stating _____ . The **null hypothesis,** denoted H_0 , states the

"assumed state of affairs" and is expressed as _____ . The **alternative**

hypothesis (also called the _____), denoted by H_1 , is the proposed substitute

to the null hypothesis and is expressed as _____ .

To test a hypothesis, we examine a _____ from the population. The claim is

either _____ by data from that random sample. If, under the assumption that

the null hypothesis is true, the observed sample proportion is *very unlikely* to have occurred by chance alone,

we _____ . Otherwise, the data does not provide us with enough

evidence to reject the null hypothesis, so we _____ .

II. Testing a Claim about a Population Proportion Using Probability: The *P*-value

The _____ associated with the observed sample is the probability of obtaining a

random sample with a proportion at least as extreme as the proportion in our random sample, given that H_0 is

true. A "very small" *P*-value tells us that it is "very unlikely" that the sample we got was obtained by chance

alone, so we should _____ . The *P*-value at which we decide to reject

the null hypothesis is called the _____ of the test, and is denoted by α .

List the steps for testing a hypothesis.

III. Inference about Two Proportions

Most statistical studies involve a comparison of two groups, usually called _____

_____. For example, when testing the effectiveness of an investigational

medication, two groups of patients are selected. The individuals in the treatment group are _____

_____; the individuals in the control group are _____.

The proportions of patients that recover in each group are compared. The goal of the study is _____

_____

_____

Let p_1 and p_2 be the true proportions of patients that recover in each group. The null hypothesis is that the

medication or treatment has _____. The alternative hypothesis is that the

treatment does have an effect: _____. The P-value is the probability that the difference

between the proportions in the two samples is due to _____. If the P-value is small

(less than the significance level of the test), we _____.

Homework Assignment

Page(s)

Exercises
